TABLE OF CONTENTS

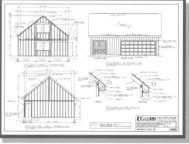

Construction Blueprints...

FULLY DETAILED BLUEPRINTS AVAILABLE

...for all project plans featured in this book.

Each Project Blueprint Plan includes the following:

- A complete list of materials
- Rafter or truss diagrams
- Framing elevations
- Fully dimensioned details
- Step-by-step instructions

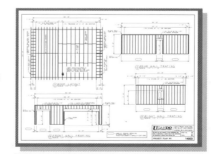

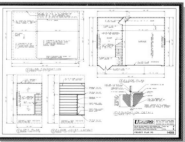

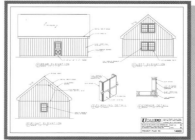

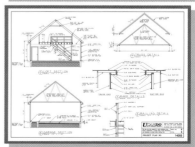

DESIGN #PB1-12001

Price Code P5

SALT BOX STORAGE SHEDS

- 3 popular sizes -
    - 8' wide x 8' deep
    - 12' wide x 8' deep
    - 16' wide x 8' deep
- Height floor to peak - 8'-2"
- Wide door opening allows easy access to your gardening tools
- Ideal storage for garden or patio equipment
- Complete list of materials
- Step-by-step instructions

DESIGN #PB1-12010

Price Code P5

GABLE STORAGE SHEDS

- 3 popular sizes -
    - 8' wide x 8' deep
    - 8' wide x 10' deep
    - 8' wide x 12' deep
- Height floor to peak - 9'-1"
- Circle-top window adds interest and light
- Complete list of materials
- Step-by-step instructions

To order plans use the form on page 83 or call toll-free 1-800-373-2646

DESIGN #PB1-12003

Price Code P5

YARD BARNS

- 3 popular sizes -
    - 10' wide x 12' deep
    - 10' wide x 16' deep
    - 10' wide x 20' deep
- Height floor to peak - 8'-4 1/2"
- Ample storage area for your lawn or garden equipment
- Complete list of materials
- Step-by-step instructions

DESIGN #PB1-12026

Price Code P5

MINI BARNS

- 4 popular sizes -
    - 8' wide x 8' deep
    - 8' wide x 10' deep
    - 8' wide x 12' deep
    - 8' wide x 16' deep
- Height floor to peak - 7'-6"
- Convenient size for storage of lawn and garden equipment
- Attractive styling perfect for any backyard
- Complete list of materials
- Step-by-step instructions

To order plans use the form on page 83 or call toll-free 1-800-373-2646

SHEDS...

DESIGN #PB1-12005

Price Code P5

GABLE STORAGE SHEDS

- 3 popular sizes -
    - 10' wide x 12' deep
    - 10' wide x 16' deep
    - 10' wide x 20' deep
- Height floor to peak - 8'-8 1/2"
- Large shed with double doors allows easy access for lawn equipment
- Complete list of materials
- Step-by-step instructions

DESIGN #PB1-12004

Price Code P5

GABLE STORAGE SHEDS

- 4 popular sizes -
    - 8' wide x 8' deep
    - 8' wide x 10' deep
    - 8' wide x 12' deep
    - 8' wide x 16' deep
- Height floor to peak - 8'-4 1/2"
- Double-door entry for easy access
- Economical and easy to build shed
- Complete list of materials
- Step-by-step instructions

To order plans use the form on page 83 or call toll-free 1-800-373-2646

DESIGN #PB1-12022

Price Code P5

YARD BARN WITH LOFT STORAGE

- Size - 10' wide x 12' deep
- Height floor to peak - 10'-7"
- Double-door entry for convenience
- Loft provides additional storage area
- Attractive styling suitable for yard
- Complete list of materials
- Step-by-step instructions

DESIGN #PB1-12002

Price Code P5

BARN STORAGE SHEDS WITH LOFT

- 3 popular sizes -
    12' wide x 12' deep
    12' wide x 16' deep
    12' wide x 20' deep
- Height floor to peak - 12'-10"
- Attractive barn-style complements any backyard or garden
- Complete list of materials
- Step-by-step instructions

SHEDS

To order plans use the form on page 83 or call toll-free 1-800-373-2646

Design #PB1-12007

Price Code P5

CONVENIENCE SHED

- Size - 16' wide x 12' deep
- Height floor to peak - 12'-4 1/2"
- Large garage-style door for lawn equipment or small boat storage
- Oversized windows brighten interior
- Complete list of materials
- Step-by-step instructions

Design #PB1-12023

Price Code P5

BARN STORAGE SHED WITH OVERHEAD DOOR

- Size - 12' wide x 16' deep
- Height floor to peak - 12'-5"
- Garage-style allows easy entry with large items
- Side windows adds light to interior
- Complete list of materials
- Step-by-step instructions

To order plans use the form on page 83 or call toll-free 1-800-373-2646

DESIGN #PB1-12019

Price Code P4

CHILDREN'S PLAYHOUSE

- Size - 6' wide x 6' deep
- Height floor to peak - 7'-2"
- Plenty of windows brighten interior
- Attractive Victorian style
- Gabled doorway and window box add interest
- Complete list of materials
- Step-by-step instructions

DESIGN #PB1-12006

Price Code P4

CHILDREN'S PLAYHOUSE

- Size - 8' wide x 8' deep
- Height floor to peak - 9'-2"
- 2' deep porch
- Attractive window boxes
- Includes operable windows
- Complete list of materials
- Step-by-step instructions

To order plans use the form on page 83 or call toll-free 1-800-373-2646

SHEDS...

DESIGN #PB1-12016

Price Code P5

STORAGE SHED WITH PLAYHOUSE LOFT

- Size - 12' wide x 12'-0" deep with 2'-8" deep balcony
- Height floor to peak - 14'-1"
- Loft above can be used as playhouse for children
- Loft features ladder for easy access
- Complete list of materials
- Step-by-step instructions

DESIGN #PB1-12025

Price Code P5

GARDEN SHED

- Size - 10' wide x 10' deep
- Height floor to peak - 11'-3 1/2"
- Wonderful complement to any backyard
- Perfect space for lawn equipment or plants and flowers
- Plenty of windows for gardening year-round
- Complete list of materials
- Step-by-step instructions

To order plans use the form on page 83 or call toll-free 1-800-373-2646

DESIGN #PB1-12021

Price Code P5

SALT BOX STORAGE SHED

- Size - 10' wide x 8' deep
- Height floor to peak - 9'-6"
- Compact yet roomy size
- Window adds light to space
- Complete list of materials
- Step-by-step instructions

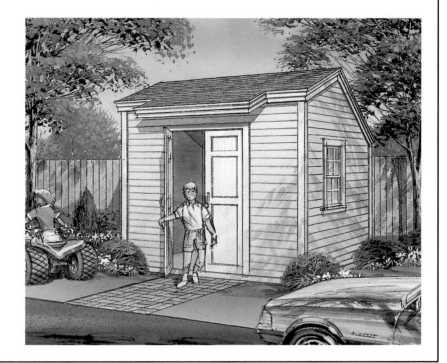

DESIGN #PB1-12024

Price Code P5

GABLE STORAGE SHED/PLAYHOUSE

- Size - 12' wide x 8' deep
- Height floor to peak - 10'-5"
- Perfect for storage or playhouse for children
- Half-door unique style
- Shutters and window box create a charming facade
- Complete list of materials
- Step-by-step instructions

SHEDS

To order plans use the form on page 83 or call toll-free 1-800-373-2646

DESIGN #PB1-12018

Price Code P5

STORAGE SHED WITH LOG BIN

- Size - 10' wide x 6' deep
- Storage area - 7'-6" wide x 6'-0" deep
- Height floor to peak - 9'-7"
- Log storage area - 2'-6" x 6'-0"
- Unique, attractive design
- Complete list of materials
- Step-by-step instructions

DESIGN #PB1-12012

Price Code P5

GABLE STORAGE SHED WITH CUPOLA

- Size - 12' wide x 10' deep
- Height floor to peak - 9'-8"
- Ideal for camp or retreat structure
- Made of cedar plywood with battens
- Complete list of materials
- Step-by-step instructions

To order plans use the form on page 83 or call toll-free 1-800-373-2646

DESIGN #PB1-12009

Price Code P5

BARN STORAGE SHEDS

- 3 popular sizes -
    - 12' wide x 8' deep
    - 12' wide x 12' deep
    - 12' wide x 16' deep
- Height floor to peak - 9'-10"
- Gambrel roof design
- Generous storage space
- Double-doors for easy access
- Complete list of materials
- Step-by-step instructions

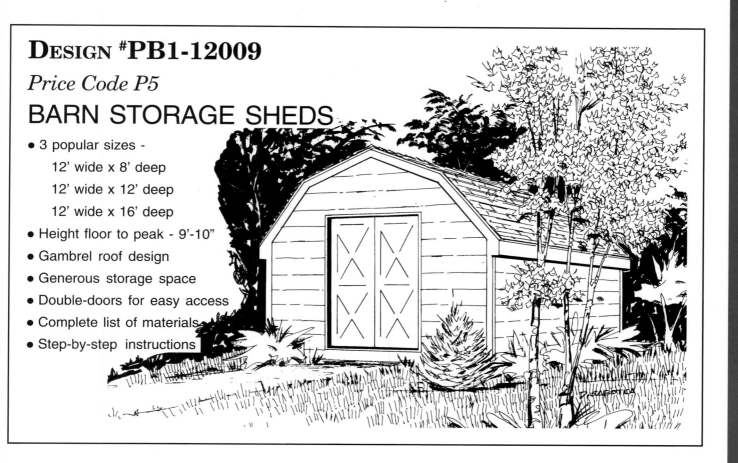

DESIGN #PB1-12015

Price Code P5

GREENHOUSE STORAGE SHED

- Size - 10' wide x 8' deep
- Height floor to peak - 7'-6"
- Plastic side panels allow for use as a greenhouse
- Handy storage size
- Complete list of materials
- Step-by-step instructions

SHEDS...

To order plans use the form on page 83 or call toll-free 1-800-373-2646

13

DESIGN #PB1-12017

Price Code P5

GARDEN SHEDS WITH CLERESTORY

- 3 popular sizes -
    - 10' wide x 10' deep
    - 12' wide x 10' deep
    - 14' wide x 10' deep
- Height floor to peak - 10'-11"
- Clerestory windows for added light
- Double-door entry for convenient access
- Complete list of materials
- Step-by-step instructions

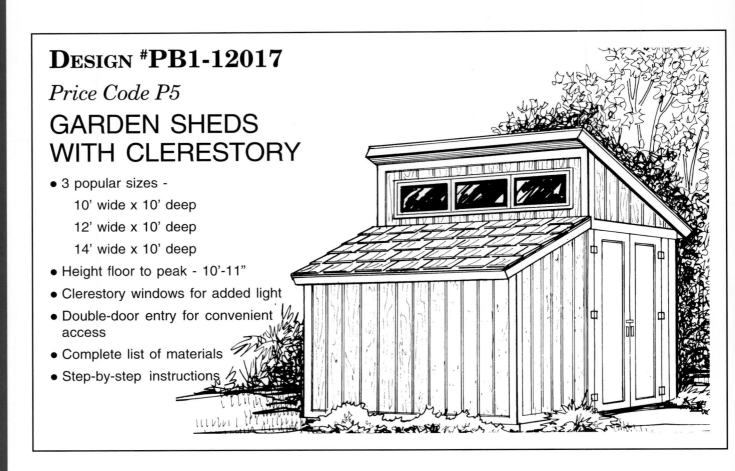

DESIGN #PB1-12020

Price Code P6

DELUXE CABANA

- Size - 11'-0" wide x 13'-6" deep
- Height floor to peak - 11'-7"
- Unique roof design with skylight
- Convenient dressing room and servicing area
- Perfect storage for poolside furniture and equipment
- Complete list of materials
- Step-by-step instructions

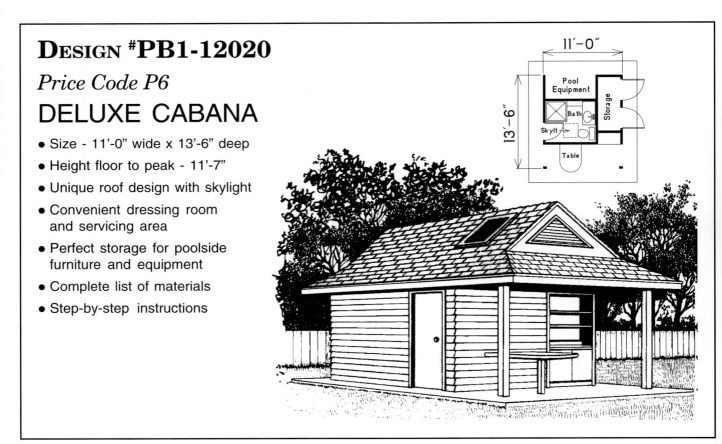

To order plans use the form on page 83 or call toll-free 1-800-373-2646

DESIGN #PB1-12014

Price Code P5

GREENHOUSE

- Size - 12' wide x 8' deep
- Height foundation to peak - 8'-3"
- An attractive addition to any yard
- Store lawn and garden tools right at hand
- Complete list of materials
- Step-by-step instructions

DESIGN #PB1-12008

Price Code P5

GARDEN SHED

- Size - 12' wide x 10' deep
- Height floor to peak - 9'-9"
- Features skylight windows for optimal plant growth
- Ample room for tool and lawn equipment storage
- Complete list of materials
- Step-by-step instructions

SHEDS...

To order plans use the form on page 83 or call toll-free 1-800-373-2646

DESIGN #PB1-12011

Price Code P5

MINI BARN STORAGE SHEDS

- 4 popular sizes -
    - 7'-3" wide x 6' deep
    - 7'-3" wide x 8' deep
    - 7'-3" wide x 10' deep
    - 7'-3" wide x 12' deep
- Height floor to peak - 9'-0"
- Attractive styling with gambrel roof
- Complete list of materials
- Step-by-step instructions

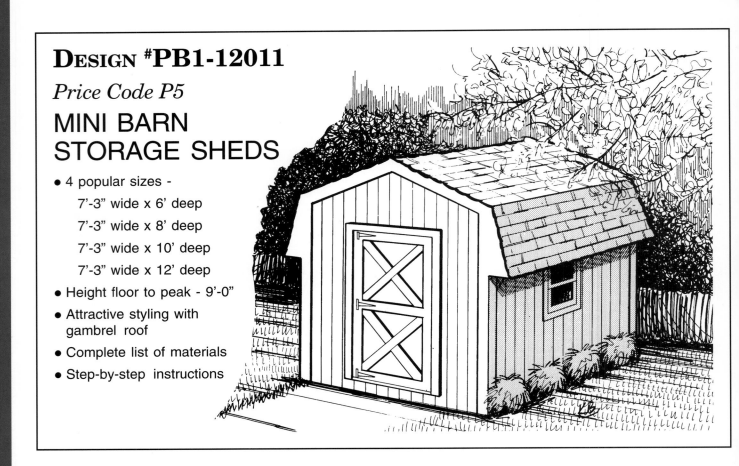

DESIGN #PB1-12013

Price Code P4

PLAYHOUSE/STORAGE SHED

- Size - 8' wide x 12' deep
- Height floor to peak - 10'-6"
- Quaint chalet design
- Ideal playhouse in summer
- Storage shed in the off season
- Complete list of materials
- Step-by-step instructions

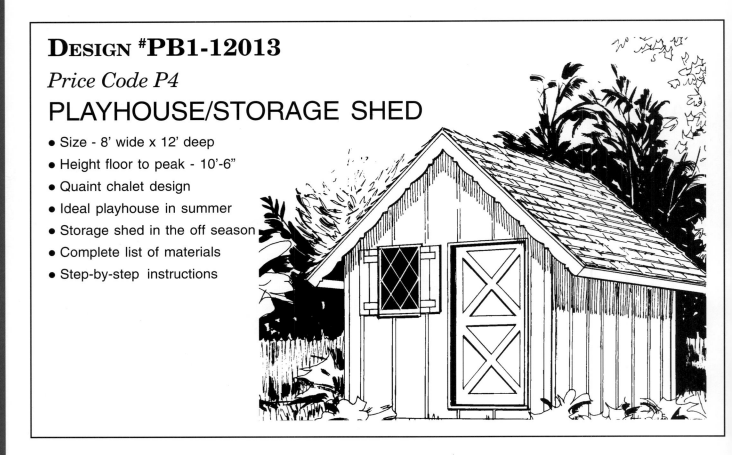

To order plans use the form on page 83 or call toll-free 1-800-373-2646

Design #PB1-11001

Price Code P2

DOGHOUSES

- Two popular sizes -
    24" x 36" x 24" high
    32" x 46" x 36" high
- Attractive gable and gambrel roof styles
- Wood cutting diagrams to help you cut cost
- Complete list of materials
- Step-by-step instructions

Design #PB1-11002

Price Code P2

BIRD HOUSES AND FEEDER

- Martin house -
    26" x 22 1/2" x 26" high
- Perfect for patio or backyard
- Designed for easy maintenance
- 3 popular sizes and styles
- Complete list of materials
- Step-by-step instructions

BACKYARD...

To order plans use the form on page 83 or call toll-free 1-800-373-2646

DESIGN #PB1-11007

Price Code P3

FENCES AND GATES - 9 DESIGNS

- Ideas for security, privacy and beauty
- From wood to chain link fencing basics
- 9 popular designs to select from
- Guides to help you estimate, buy and build
- Complete list of materials
- Step-by-step instructions

DESIGN #PB1-11004

Price Code P3

ADIRONDACK CHAIR

- Size - 66" x 27" x 40" high
- A project that's very practical and unique
- Two piece set
- Sturdy construction
- Complete list of materials
- Step-by-step instructions

To order plans use the form on page 83 or call toll-free 1-800-373-2646

DESIGN #PB1-11015

Price Code P3

ALL PURPOSE BENCH

- Enhance your garden, patio or deck
- Complements any setting
- Size - 72" x 20" x 36" high
- Complete list of materials
- Step-by-step instructions

DESIGN #PB1-11013

Price Code P3

LEISURE BENCH WITH TABLE

- Bench size - 60" x 20" x 36" high
- Table size - 48" x 20" x 18" high
- Enhance your garden, patio or deck
- Complements any setting
- Complete list of materials
- Step-by-step instructions

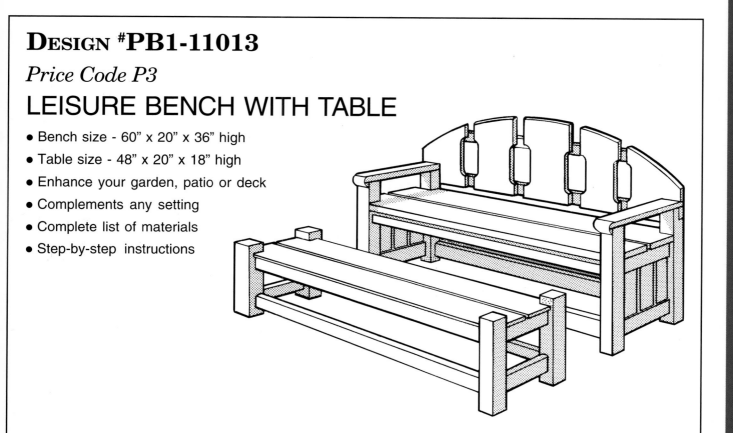

To order plans use the form on page 83 or call toll-free 1-800-373-2646

DESIGN #PB1-11006

Price Code P3

PATIO FURNITURE - 3 PIECE SET

- Lounge seat - 63" x 32" x 31" high
- Ideal for patio or deck
- Convenient for outdoor entertaining
- Complete list of materials
- Step-by-step instructions

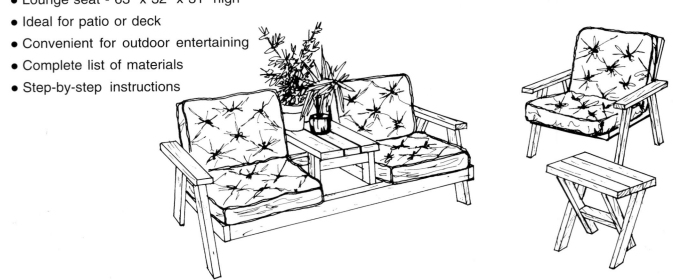

DESIGN #PB1-11010

Price Code P2

PVC OUTDOOR FURNITURE

- Perfect for patio or deck
- Designed for easy maintenance
- Very practical and unique
- Simple construction using plastic pipe to build this chair and chaise lounge
- Complete list of materials
- Step-by-step instructions

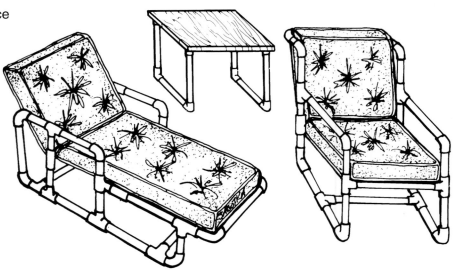

To order plans use the form on page 83 or call toll-free 1-800-373-2646

DESIGN #PB1-11005

Price Code P3

PICNIC KING BENCH AND TABLE

- Table size - 6'-0" x 5'-2"
- Bench size - 6'-0" x 2'-7"
- Converts from table to bench
- Wood cutting diagrams to help you cut cost
- Sturdy construction
- Complete list of materials
- Step-by-step instructions

DESIGN #PB1-11003

Price Code P3

PICNIC TABLES

- Two popular styles -
  Rectangle - 72" x 60" x 30" high
  Octagon - 56" x 56" x 30" high
- Ideal for outdoor entertaining and backyard barbeques
- Complete list of materials
- Step-by-step instructions

To order plans use the form on page 83 or call toll-free 1-800-373-2646

BACKYARD...

DESIGN #PB1-11012

Price Code P3

GARDEN SWING WITH CANOPY

- Canopy size - 12'-0" x 5'-0" x 7'-6" high
- Bench size - 6'-0" long
- Attractive design features a sun screen canopy
- Perfect for enjoying the outdoors in style
- Complete list of materials
- Step-by-step instructions

DESIGN #PB1-11016

Price Code P3

PORCH SWING

- Ideal for attaching to porch or any outside structure
- Attractive and sturdy design
- Size - 72" x 24" x 26" high
- Complete list of materials
- Step-by-step instructions

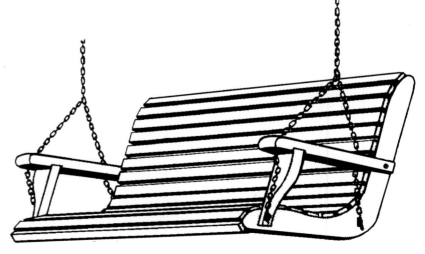

To order plans use the form on page 83 or call toll-free 1-800-373-2646

BACKYARD...

DESIGN #PB1-11011

Price Code P3

JUNGLE GYM - SWING SET

- Size 13' x 13' x 10' high
- Designed with versatility in mind
- Simple construction for easy assembly
- Plenty of outside enjoyment for everyone in the family
- Complete list of materials
- Step-by-step instructions

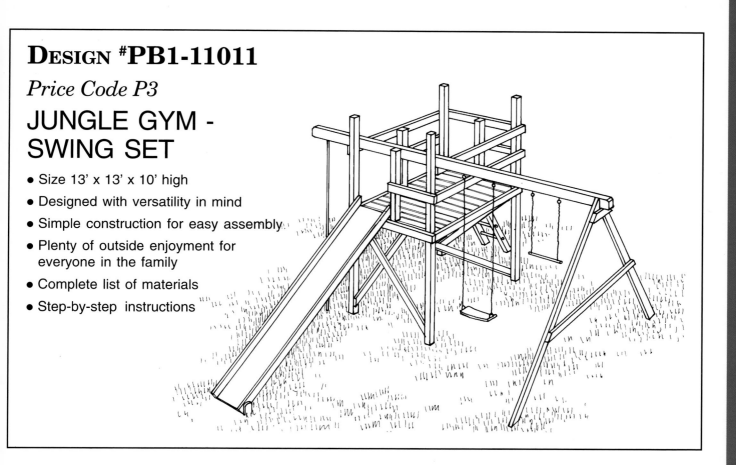

DESIGN #PB1-11009

Price Code P3

JUNGLE GYM - MULTI-LEVEL

- Size - 8' x 8' x 10' high
- Multi-level platforms create a unique play structure
- Outside fun for children of all ages
- Complete list of materials
- Step-by-step instructions

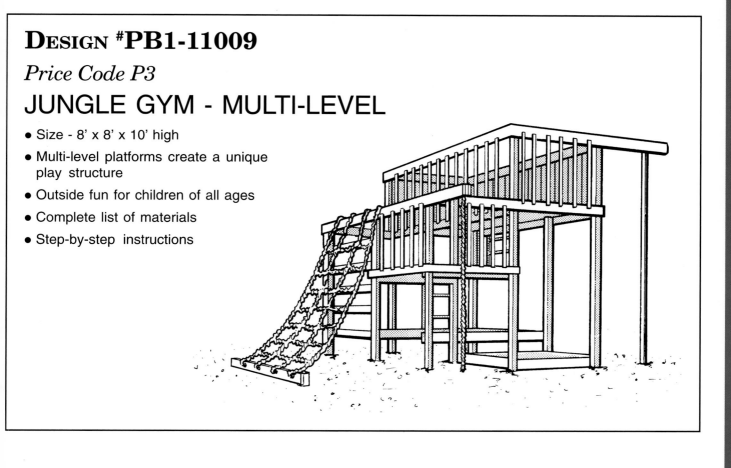

BACKYARD...

To order plans use the form on page 83 or call toll-free 1-800-373-2646

DESIGN #PB1-11008

Price Code P2

TWO CUPOLAS

- Sizes - Plan A: 30" x 30" x 40" high
  - Plan B: 33" x 33" x 60" high
- A decorative finishing touch for any structure
- Easy to build
- Complete list of materials
- Step-by-step instructions

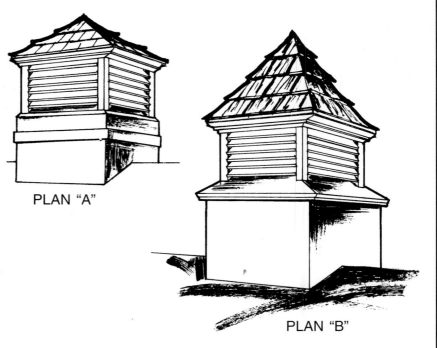

PLAN "A"

PLAN "B"

DESIGN #PB1-11014

Price Code P2

COMPOSTER WITH MANUAL

- Size - 48" x 48" x 48"
- Practical and functional waste storage
- Sturdy design
- Complete list of materials
- Step-by-step instructions

To order plans use the form on page 83 or call toll-free 1-800-373-2646

DESIGN #PB1-14018

Price Code P7

2 CAR GARAGE WITH STORAGE

- Size - 24' x 26'
- Plenty of storage for yard equipment
- Convenient side entry
- Complete list of materials
- Step-by-step instructions

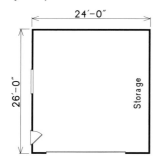

DESIGN #PB1-14001

Price Code P8

2 CAR GARAGE WITH LOFT - GAMBREL ROOF

- Size - 22' x 26'
- Spacious loft above
- Complete list of materials
- Step-by-step instructions

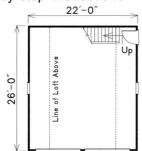

To order plans use the form on page 83 or call toll-free 1-800-373-2646

GARAGES...

25

DESIGN #PB1-14030

Price Code P6

1 CAR GARAGES

- 4 popular sizes -
  - 14' x 22' 14' x 24'
  - 16' x 22' 16' x 24'
- Sturdy, attractive design
- Complete list of materials
- Step-by-step instructions

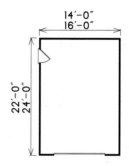

DESIGN #PB1-14004

Price Code P7

TWO CAR GARAGE

- Size - 24' x 24'
- Features two single garage doors
- Side-entry is efficient and well-designed
- Complete list of materials
- Step-by-step instructions

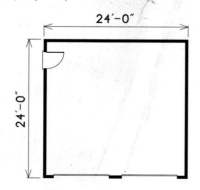

To order plans use the form on page 83 or call toll-free 1-800-373-2646

GARAGES...

DESIGN #PB1-14045

Price Code P8

1 CAR GARAGE WITH LOFT - GAMBREL ROOF

- Size - 16' x 24'
- Ideal loft perfect for children's play area or workshop
- Handy side door
- Complete list of materials
- Step-by-step instructions

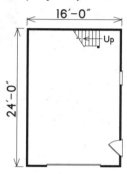

DESIGN #PB1-14047

Price Code P6

2 CAR CARPORT WITH STORAGE

- Size - 24' x 24'
- Unique design allows cars to enter from the front or the side of carport
- Deep storage space for long or tall items
- Complete list of materials
- Step-by-step instructions

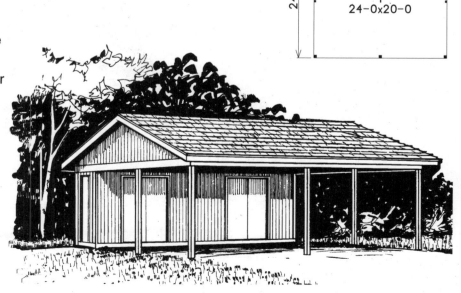

DESIGN #PB1-14034

Price Code P6

2 CAR ECONOMY GARAGE

- Size - 20' x 20'
- Convenient side-entry
- Complete list of materials
- Step-by-step instructions

DESIGN #PB1-14013

Price Code P6

2 CAR GARAGE - HIP ROOF

- Size - 22' x 22'
- Handsome styling, ideal with many home types
- Practical side-door entry
- Complete list of materials
- Step-by-step instructions

To order plans use the form on page 83 or call toll-free 1-800-373-2646

DESIGN #PB1-14036

Price Code P6

2 CAR ECONOMY GARAGE - HIP ROOF

- Size - 20' x 20'
- Attractive hip roof design l
- Extended roof over garage door protects from the weather
- Complete list of materials
- Step-by-step instructions

DESIGN #PB1-14037

Price Code P6

2 CAR ECONOMY GARAGE

- Size - 20' x 20'
- Practical and functional
- Convenient side-entry
- Complete list of materials
- Step-by-step instructions

To order plans use the form on page 83 or call toll-free 1-800-373-2646

DESIGN #PB1-14015

Price Code P6

2 CAR GARAGE - ATTACHED OR DETACHED

- Size - 24' x 22'
- Practical styling
- Wonderful versatility with this design
- Complete list of materials
- Step-by-step instructions

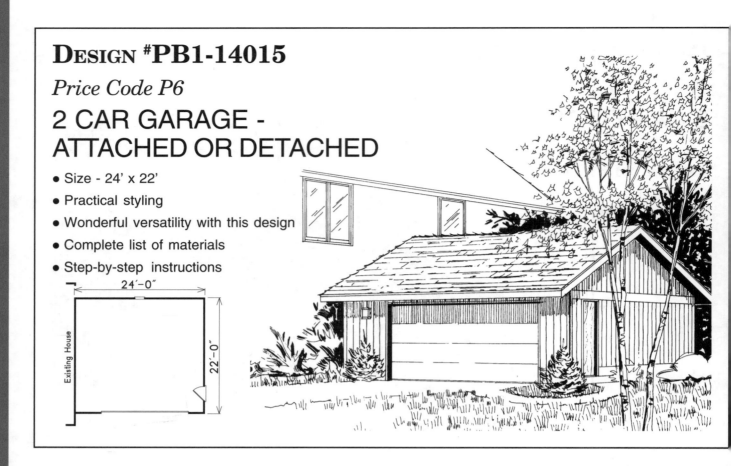

DESIGN #PB1-14032

Price Code P6

2 CAR GARAGE - ATTACHED OR DETACHED

- Size - 22' x 24'
- Convenient service door on front
- Traditionally styled
- Complete list of materials
- Step-by-step instructions

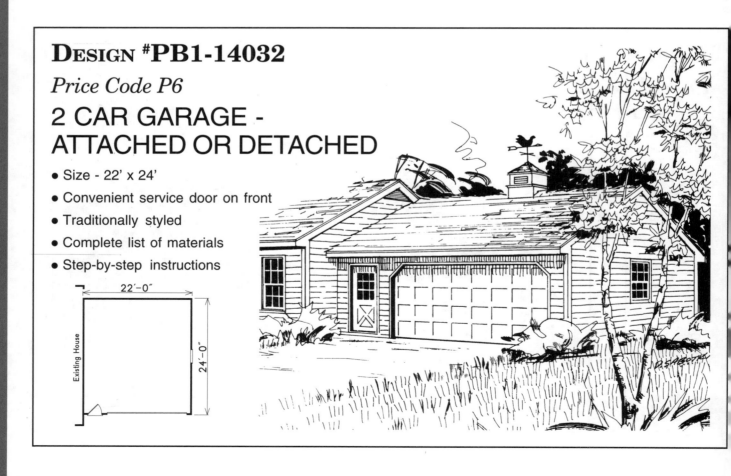

To order plans use the form on page 83 or call toll-free 1-800-373-2646

GARAGES...

DESIGN #PB1-14009

Price Code P7

2 CAR GARAGE WITH STORAGE-REVERSE GABLE

- Size - 24' x 24'
- Windows on two sides
- Extra space perfect for storage
- Complete list of materials
- Step-by-step instructions

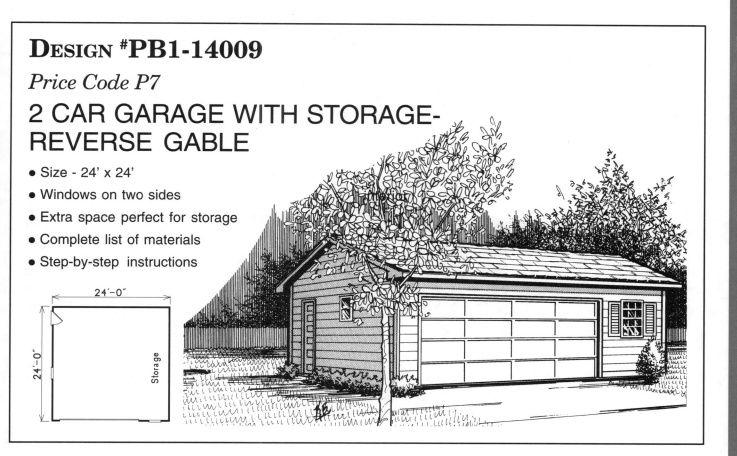

DESIGN #PB1-14007

Price Code P7

2 CAR GARAGE - REVERSE GABLE

- Size - 24' x 24'
- Oversized, appealing design
- Side door is a handy feature
- Complete list of materials

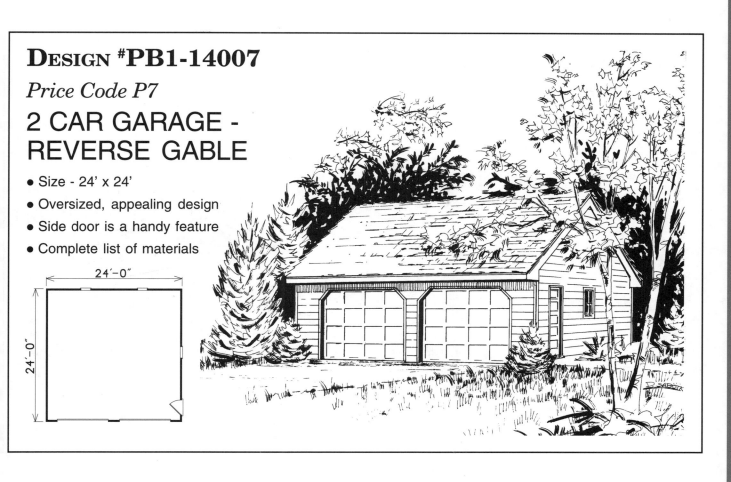

DESIGN #PB1-14019

Price Code P7

2 CAR GARAGE - VICTORIAN

- Size - 24' x 24'
- Accented with Victorian details
- Functional side entry
- Complete list of materials
- Step-by-step instructions

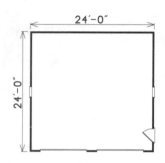

DESIGN #PB1-14024

Price Code P7

2 CAR GARAGE - WESTERN STYLE

- Size - 24' x 24'
- Appealing style with many homes
- Side-entry door and window are functional extras
- Complete list of materials
- Step-by-step instructions

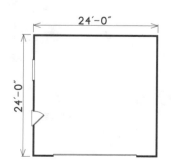

To order plans use the form on page 83 or call toll-free 1-800-373-2646

DESIGN #PB1-14022

Price Code P6

2 CAR GARAGE

- Size - 22' x 24'
- Attractive style for any home type
- Appealing side entry
- Complete list of materials
- Step-by-step instructions

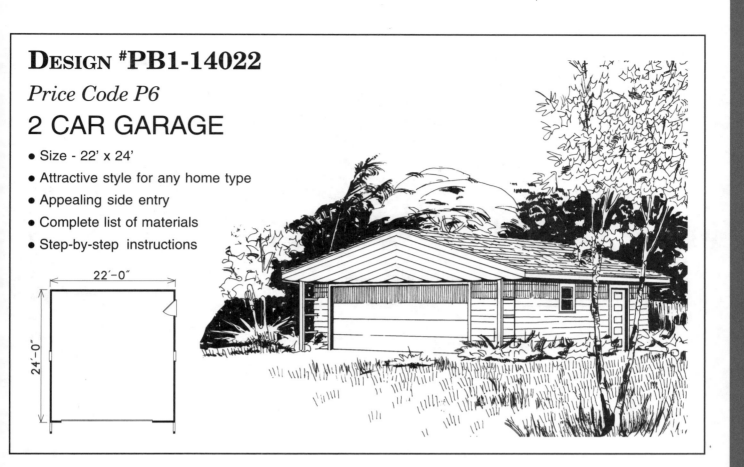

22'-0"

24'-0"

DESIGN #PB1-14006

Price Code P6

1 CAR GARAGE

- Size - 14' x 22'
- Side window enhances exterior
- Side-entry is convenient and useful
- Complete list of materials
- Step-by-step instructions

14'-0"

22'-0"

GARAGES...

To order plans use the form on page 83 or call toll-free 1-800-373-2646

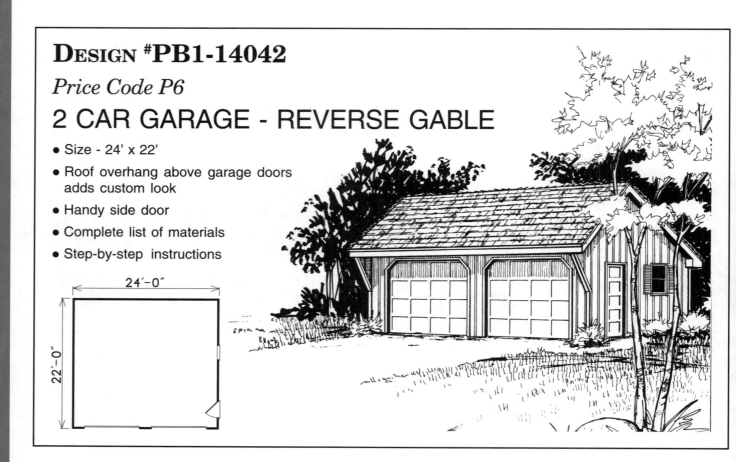

DESIGN #PB1-14042

Price Code P6

2 CAR GARAGE - REVERSE GABLE

- Size - 24' x 22'
- Roof overhang above garage doors adds custom look
- Handy side door
- Complete list of materials
- Step-by-step instructions

24'-0"

22'-0"

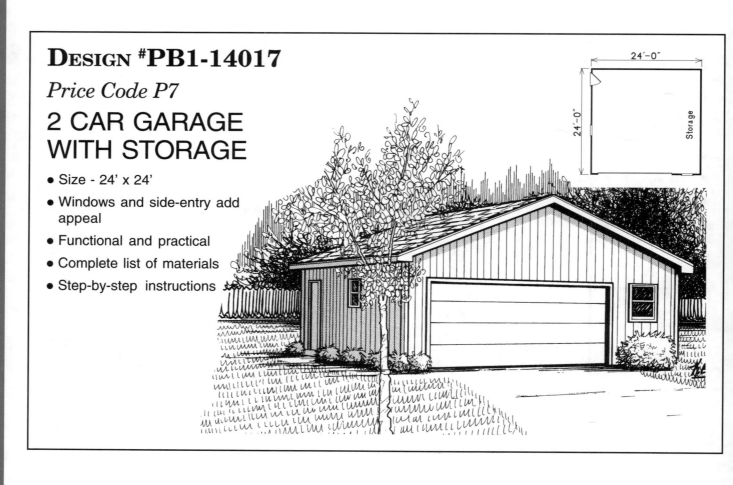

DESIGN #PB1-14017

Price Code P7

2 CAR GARAGE WITH STORAGE

- Size - 24' x 24'
- Windows and side-entry add appeal
- Functional and practical
- Complete list of materials
- Step-by-step instructions

24'-0"

24'-0"

Storage

GARAGES...

To order plans use the form on page 83 or call toll-free 1-800-373-2646

DESIGN #PB1-14014

Price Code P6

2 CAR GARAGE

- Size - 22' x 22'
- Useful side-entry door
- Perfect for tractor or lawn equipment
- Complete list of materials
- Step-by-step instructions

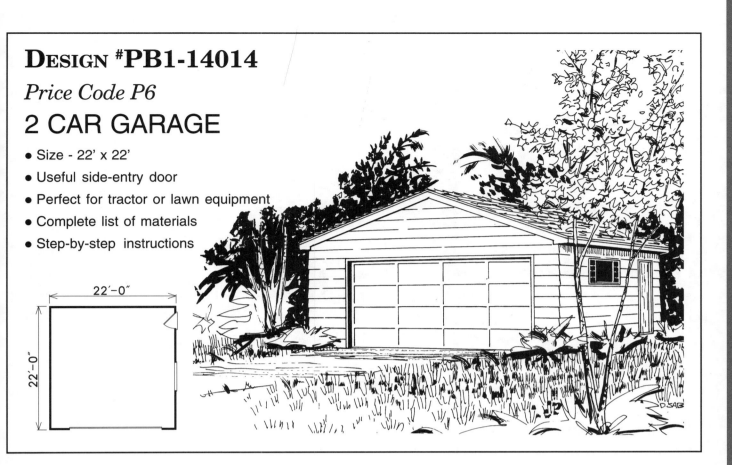

DESIGN #PB1-14010

Price Code P7

2 CAR GARAGE WITH STORAGE-HIP ROOF

- Size - 24' x 24'
- Attractive and unique style
- Side-entry provides easy access
- Complete list of materials
- Step-by-step instructions

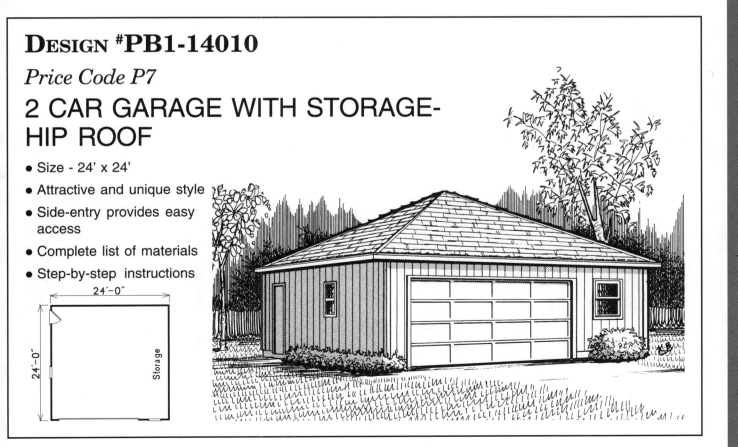

To order plans use the form on page 83 or call toll-free 1-800-373-2646

GARAGES...

35

Design #PB1-14020

Price Code P7

2 CAR GARAGE WITH 8' HIGH DOOR

- Size - 24' x 26'
- Practical and appealing
- Side window adds light
- Complete list of materials
- Step-by-step instructions

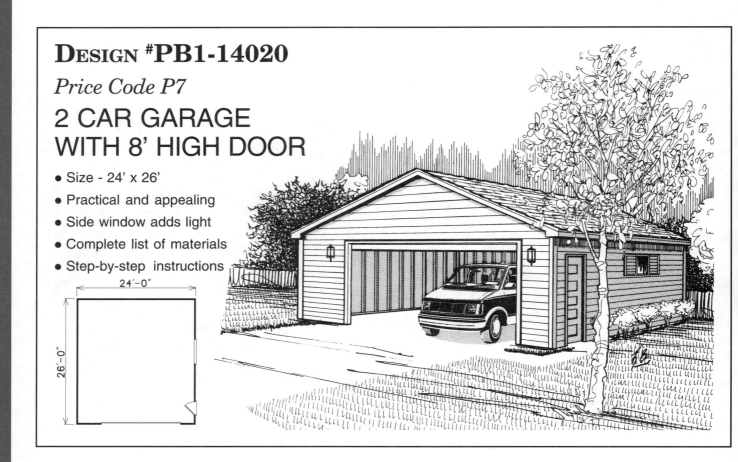

24'-0"

26'-0"

Design #PB1-14035

Price Code P6

2 CAR GARAGE - REVERSE GABLE

- Size - 22' x 24'
- Two 9' x 7' overhead doors
- Complete list of materials
- Step-by-step instructions

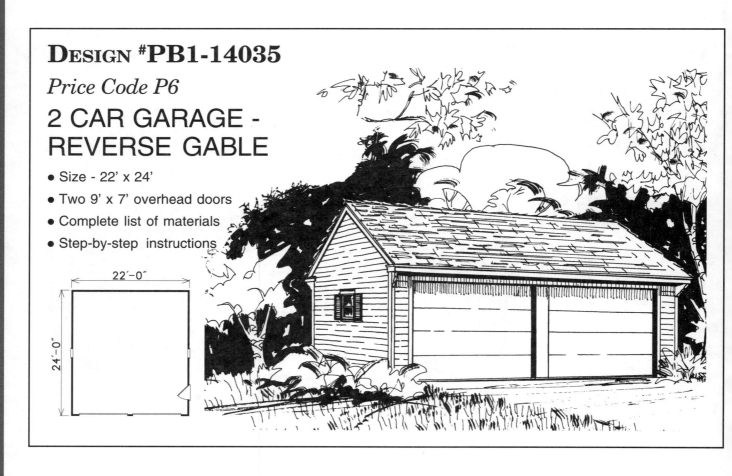

22'-0"

24'-0"

To order plans use the form on page 83 or call toll-free 1-800-373-2646

GARAGES...

DESIGN #PB1-14038

Price Code P6

2 CAR GARAGE - REVERSE GABLE

- Size - 24' x 22'
- Roof overhang above garage door to protect from the weather
- Handy side door
- Complete list of materials
- Step-by-step instructions

DESIGN #PB1-14039

Price Code P7

2 CAR GARAGE WITH STORAGE

- Size - 26' x 22'
- Provides two separate lockable storage compartments, one of which is accessible from the outside
- Helpful addition to your home
- Complete list of materials
- Step-by-step instructions

To order plans use the form on page 83 or call toll-free 1-800-373-2646

GARAGES...

37

Design #PB1-14023

Price Code P6

1 CAR GARAGE - WESTERN STYLE

- Size - 14' x 22'
- Compact size, perfect for smaller lots
- Efficient side door provides easy access
- Complete list of materials
- Step-by-step instructions

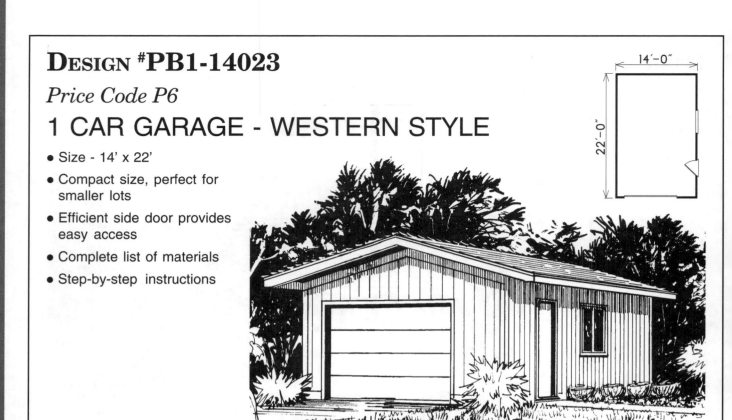

Design #PB1-14028

Price Code P7

2 CAR GARAGE - WESTERN STYLE/REVERSE GABLE

- Size - 24' x 24'
- Easy, functional design
- Two single garage doors
- Complete list of materials
- Step-by-step instructions

To order plans use the form on page 83 or call toll-free 1-800-373-2646

GARAGES...

DESIGN #PB1-14040

Price Code P7

2 CAR GARAGE WITH STORAGE

- Size - 26' x 22'
- Attractive salt-box style
- Includes additional storage
- Complete list of materials
- Step-by-step instructions

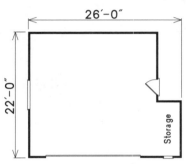

26'-0"

22'-0"

Storage

DESIGN #PB1-14033

Price Code P7

2 CAR GARAGE - GAMBREL ROOF

- Size - 24' x 24'
- Attractive addition to any home
- Complete list of materials
- Step-by-step instructions

24'-0"

24'-0"

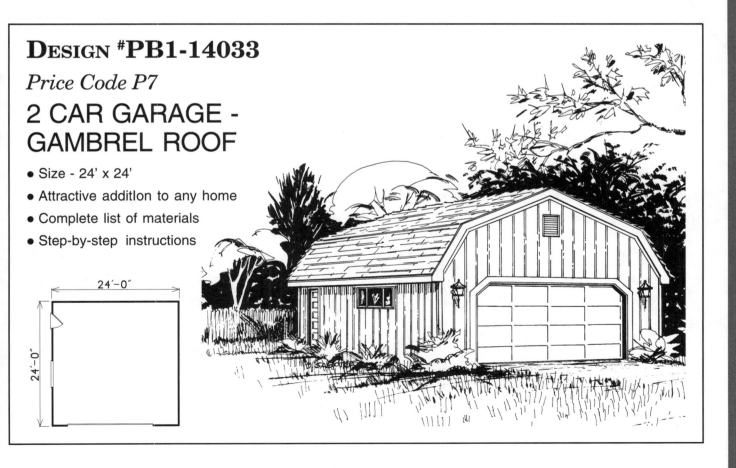

To order plans use the form on page 83 or call toll-free 1-800-373-2646

DESIGN #PB1-14031

Price Code P7

2 CAR GARAGE WITH GREENHOUSE

- Size - 30' x 24'
- Unique design allows year-round gardening
- Additional space perfect for storing lawn equipment
- Complete list of materials
- Step-by-step instructions

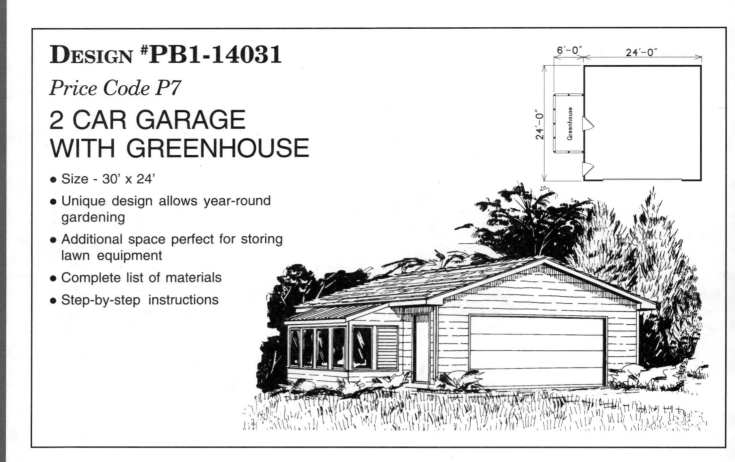

DESIGN #PB1-14011

Price Code P7

1 CAR GARAGE WITH COVERED PORCH

- Size - 24' x 22'
- Roomy garage has space for storage
- Distinctive covered porch provides perfect area for entertaining
- Complete list of materials
- Step-by-step instructions

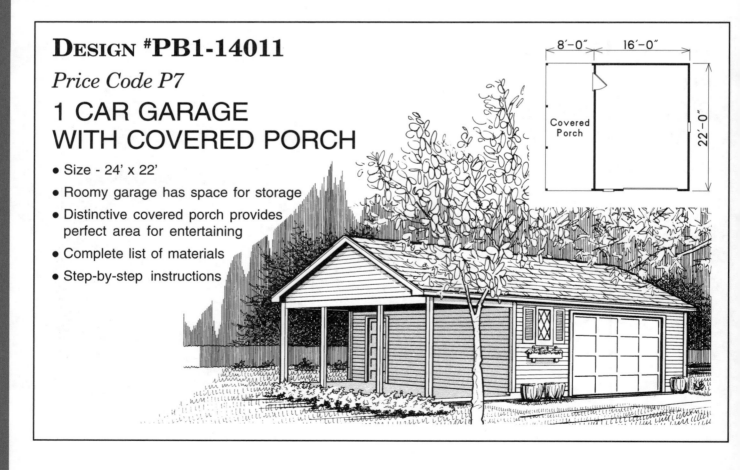

To order plans use the form on page 83 or call toll-free 1-800-373-2646

Design #PB1-14041

Price Code P8

2 CAR GARAGE WITH LOFT

- Size - 28' x 24'
- Charming dormers add character
- Handy side door accessing stairs to loft
- Complete list of materials
- Step-by-step instructions

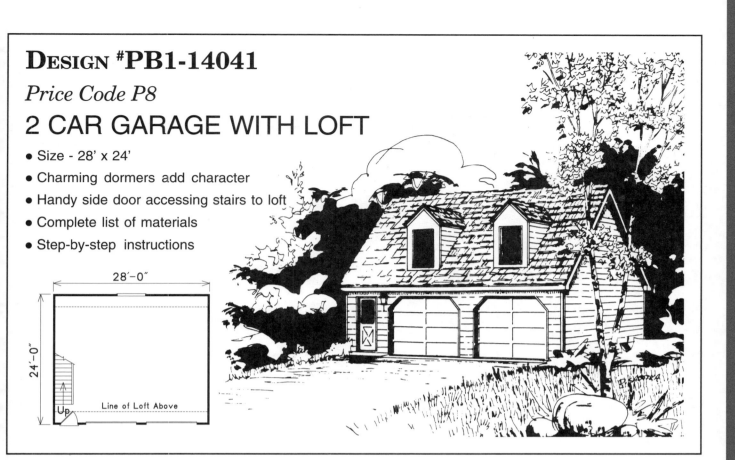

28'-0"

24'-0"

Line of Loft Above

Up

Design #PB1-14005

Price Code P8

2 CAR GARAGE WITH WORKSHOP AND LOFT

- Size - 32' x 24'
- Loft space perfect for studio or home office
- Plenty of storage space for workshop or hobby center
- Complete list of materials
- Step-by-step instructions

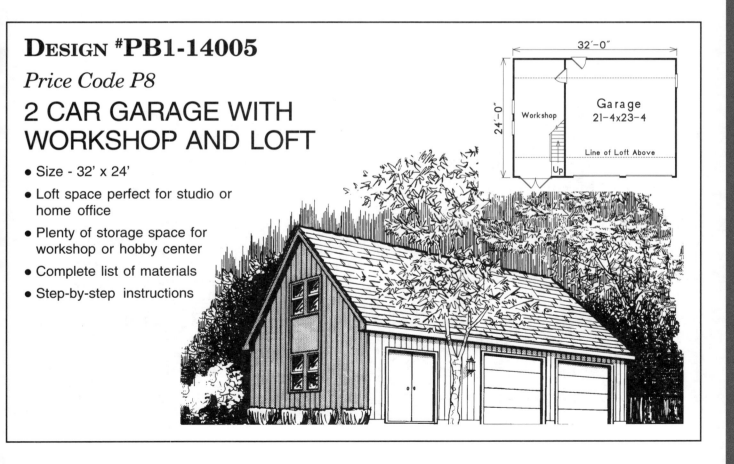

32'-0"

24'-0"

Workshop

Garage
21-4x23-4

Line of Loft Above

Up

Design #PB1-14003

Price Code P8

TWO CAR GARAGE WITH WORKSHOP AND PARTIAL LOFT

- Size - 32' x 24'
- Convenient loft above workshop for work space or storage
- Unique double-door entry is functional and practical
- Complete list of materials
- Step-by-step instructions

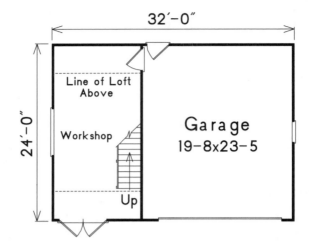

DESIGN #PB1-14002

Price Code P8

2 CAR GARAGE WITH LOFT

- Size - 28' x 24'
- Appealing dormers add spaciousness and light
- Attractive Cape Cod style
- Complete list of materials
- Step-by-step instructions

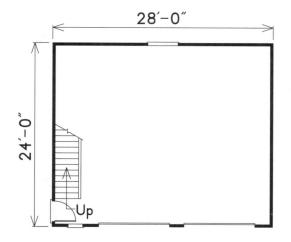

To order plans use the form on page 83 or call toll-free 1-800-373-2646

DESIGN #PB1-14016

Price Code P8

2 CAR GARAGE WITH LOFT

- Size - 26' x 24'
- Loft provides extra storage area or workshop space
- Clerestory windows brighten inside
- Complete list of materials
- Step-by-step instructions

26'-0"

24'-0"

Line of Loft Above

Up

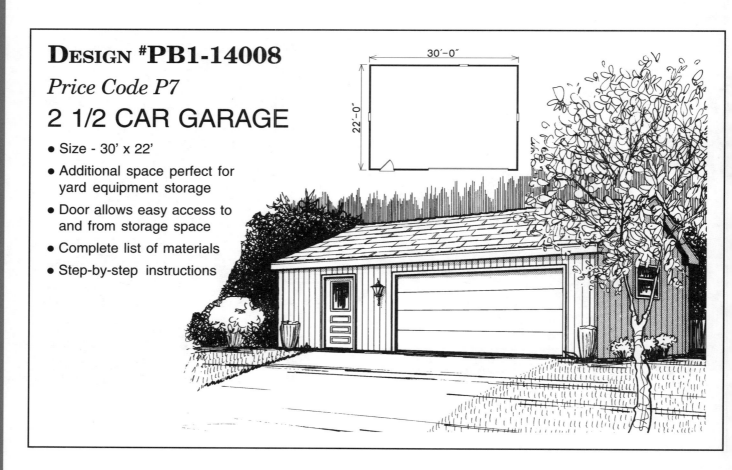

DESIGN #PB1-14008

Price Code P7

2 1/2 CAR GARAGE

- Size - 30' x 22'
- Additional space perfect for yard equipment storage
- Door allows easy access to and from storage space
- Complete list of materials
- Step-by-step instructions

30'-0"

22'-0"

GARAGES...

To order plans use the form on page 83 or call toll-free 1-800-373-2646

DESIGN #PB1-14025

Price Code P7

2 1/2 CAR GARAGE - WESTERN STYLE

- Size - 30' x 24'
- Plenty of storage space
- Additional space perfect for workshop
- Complete list of materials
- Step-by-step instructions

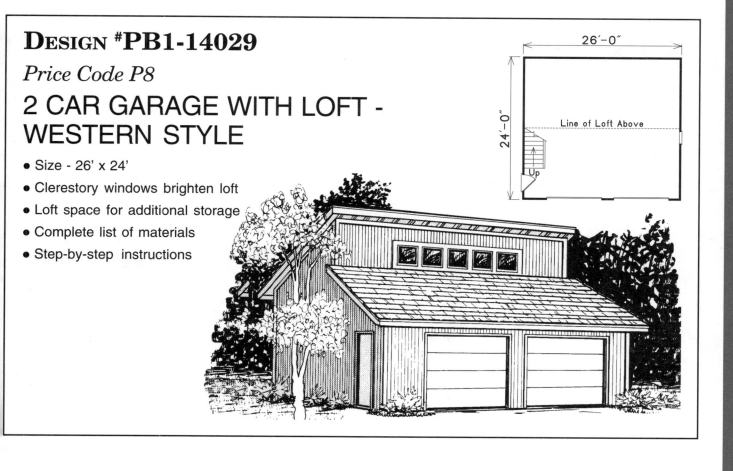

30'-0"

24'-0"

DESIGN #PB1-14029

Price Code P8

2 CAR GARAGE WITH LOFT - WESTERN STYLE

- Size - 26' x 24'
- Clerestory windows brighten loft
- Loft space for additional storage
- Complete list of materials
- Step-by-step instructions

26'-0"

24'-0"

Line of Loft Above

Up

To order plans use the form on page 83 or call toll-free 1-800-373-2646

DESIGN #PB1-14043

Price Code P7

2 1/2 CAR GARAGE/ROADSIDE STAND

- Size - 32' x 30'
- Excellent for displaying, selling and storing fresh produce
- 6' cantilevered front overhang
- Complete list of materials
- Step-by-step instructions

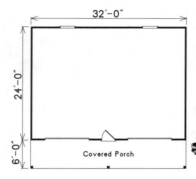

32'-0"

24'-0"

6'-0"

Covered Porch

DESIGN #PB1-14044

Price Code P7

3 CAR GARAGE/WORKSHOP

- Size - 24' x 36'
- Oversized for storage
- Ideal size for workshop or maintenance building
- Complete list of materials
- Step-by-step instructions

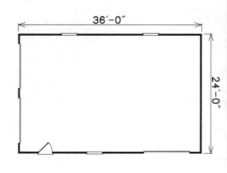

36'-0"

24'-0"

To order plans use the form on page 83 or call toll-free 1-800-373-2646

GARAGES...

DESIGN #PB1-14048

Price Code P7

3 CAR GARAGE

- Size - 40' x 24'
- Oversized with plenty of room for storage
- Side door for easy access
- Complete list of materials
- Step-by-step instructions

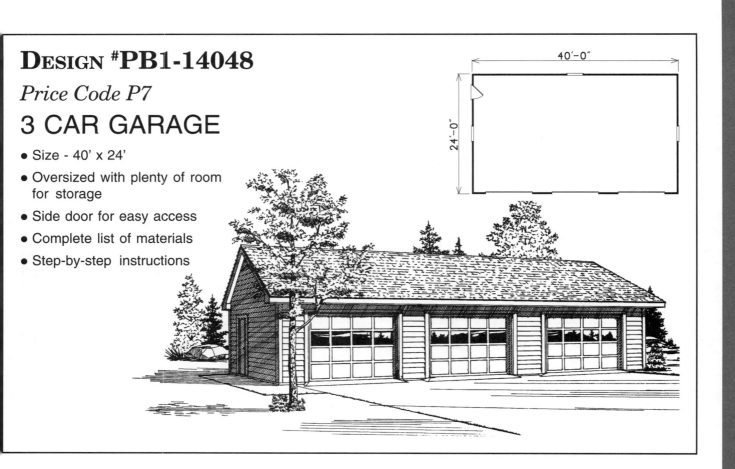

DESIGN #PB1-14021

Price Code P7

3 CAR GARAGE WITH WORKSHOP

- Size - 32' x 28'
- Handy workshop space for hobbies
- Side-entry door provides easy access
- Complete list of materials
- Step-by-step instructions

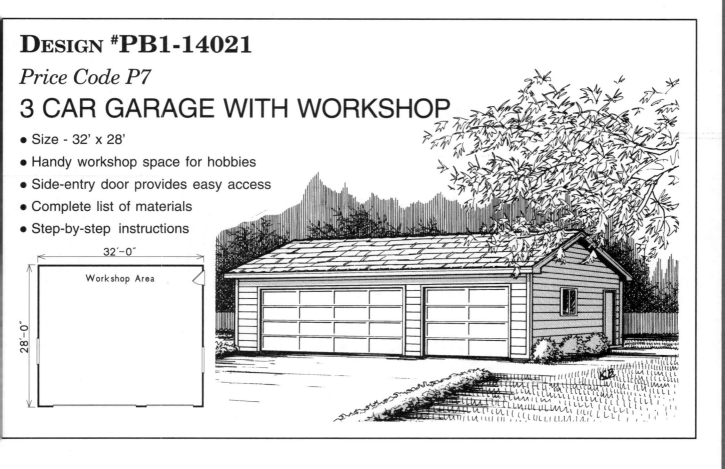

To order plans use the form on page 83 or call toll-free 1-800-373-2646

GARAGES...

47

DESIGN #PB1-14012

Price Code P7

3 CAR GARAGE

- Size - 32' x 22'
- Side-entry for easy access
- Perfect style with many types of homes
- Complete list of materials
- Step-by-step instructions

DESIGN #PB1-14046

Price Code P8

3 CAR GARAGE WITH LOFT

- Size - 36' x 24'
- Third stall in garage perfect for boat storage
- Generous loft space for storage or studio
- Complete list of materials
- Step-by-step instructions

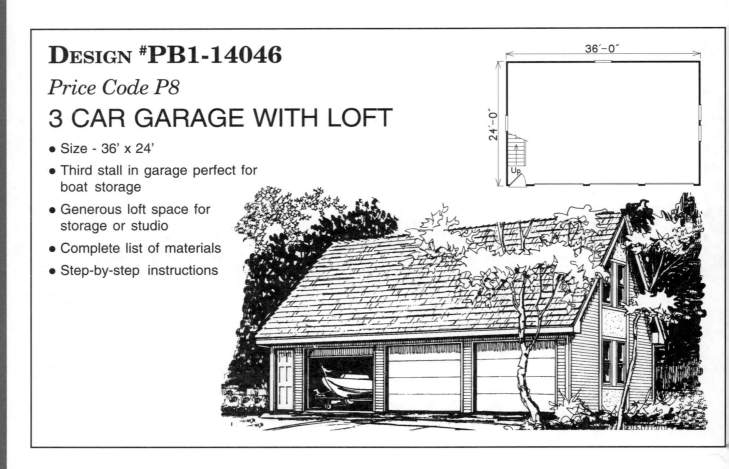

To order plans use the form on page 83 or call toll-free 1-800-373-2646

DESIGN #PB1-14027

Price Code P8

3 CAR GARAGE WITH LOFT - WESTERN STYLE

- Size - 32' x 24'
- Large side windows draw in light
- Loft space perfect for studio or home office
- Complete list of materials
- Step-by-step instructions

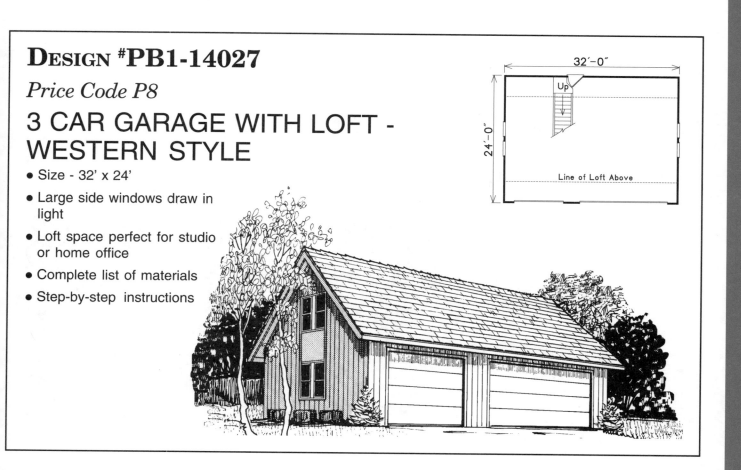

32'-0"

24'-0"

Up

Line of Loft Above

DESIGN #PB1-14026

Price Code P7

3 CAR GARAGE

- Size - 30' x 24'
- Highly functional design
- Handy side-entry door
- Complete list of materials
- Step-by-step instructions

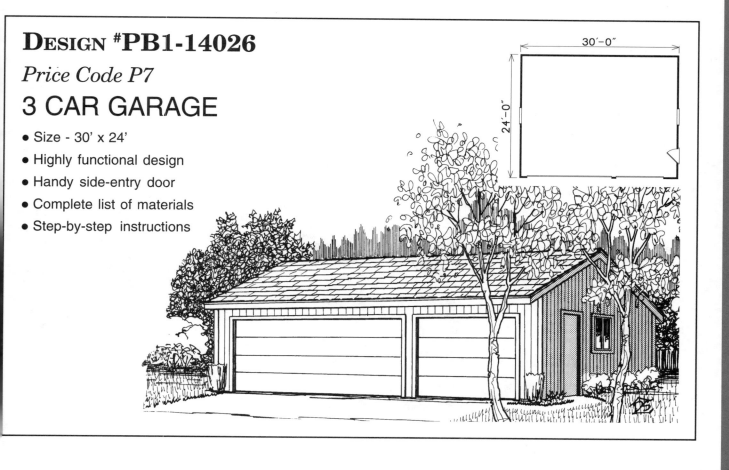

30'-0"

24'-0"

To order plans use the form on page 83 or call toll-free 1-800-373-2646

GARAGES...

DESIGN #PB1-15011

Price Code P9

2 CAR GARAGE APARTMENT WITH INTERIOR ENTRANCE

- Size - 28' x 26'
- Building height 21'-4"
- Contemporary style ideal for extended family or income property
- 1 bedroom, 1 bath
- Complete list of materials
- Step-by-step instructions

DESIGN #PB1-15015

Price Code P9

2 CAR GARAGE APARTMENT - TUDOR

- Size - 28' x 28'
- 784 square feet
- Outside covered stairs shelter from the elements
- 1 bedroom, 1 bath
- Complete list of materials
- Step-by-step instructions

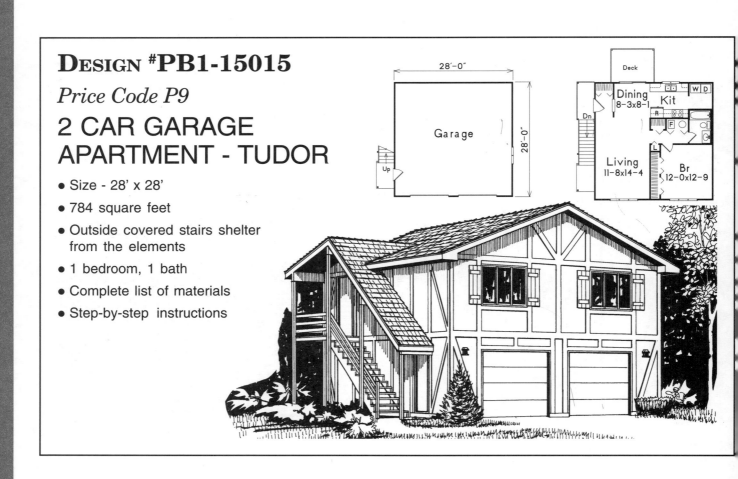

To order plans use the form on page 83 or call toll-free 1-800-373-2646

BUILDINGS...

DESIGN #PB1-15029

Price Code P9

2 CAR GARAGE - GAMBREL ROOF

- Size - 24' x 26'
- Comfortable colonial-styling
- Simple yet spacious studio design
- Large windows warm inside
- Complete list of materials
- Step-by-step instructions

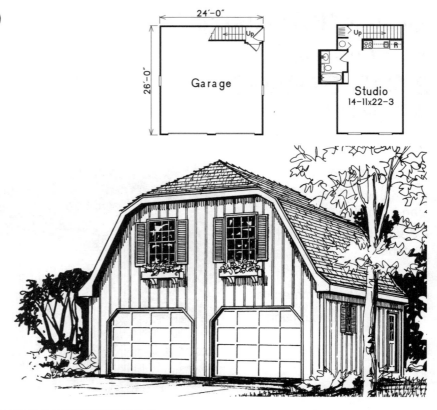

DESIGN #PB1-15030

Price Code P9

2 CAR GARAGE APARTMENT WITH EXTERIOR ENTRANCE

- Size - 24' x 24'
- Loft has roomy kitchen and dining area
- Private side exterior entrance
- Style complements many types of homes
- Complete list of materials
- Step-by-step instructions

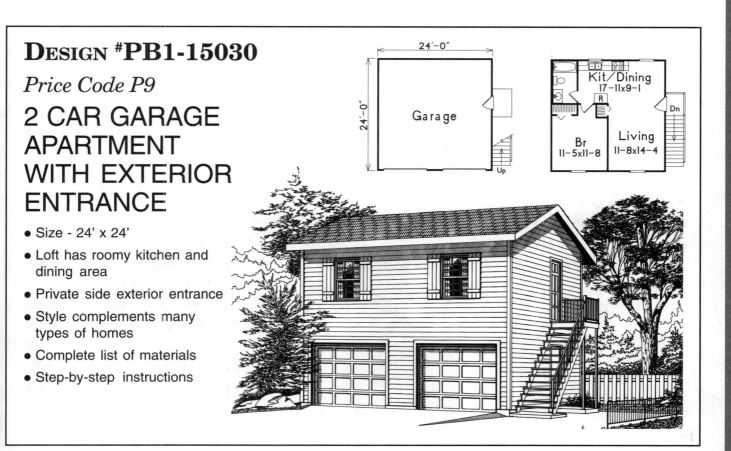

To order plans use the form on page 83 or call toll-free 1-800-373-2646

BUILDINGS...

51

DESIGN #PB1-15020

Price Code P9

2 CAR GARAGE APARTMENT - WESTERN STYLE

- Size - 28' x 28'
- Open living area spacious and functional
- Space for utilities off the kitchen
- 1 bedroom, 1 bath
- Complete list of materials
- Step-by-step instructions

28'-0"

28'-0"

Garage

Up

Deck

Dining
8-3x8-1

Kit

W D

Dn

R

F

Living
11-8x14-4

Br
12-0x12-9

DESIGN #PB1-15028

Price Code P9

2 CAR GARAGE APARTMENT - CAPE COD

- Size - 28' x 24'
- Building height - 22'-0"
- Charming dormers add appeal to this design
- Comfortable open living area
- Complete list of materials
- Step-by-step instructions

28'-0"

24'-0"

Garage

Up

Up

Studio
18-2x18x4

R

BUILDINGS...

DESIGN #PB1-15026

Price Code P9

2 CAR GARAGE APARTMENT - GAMBREL ROOF

- Size - 28' x 24'
- Building height - 21'-4"
- Charming dutch colonial style
- Spacious studio provides extra storage space
- Complete list of materials
- Step-by-step instructions

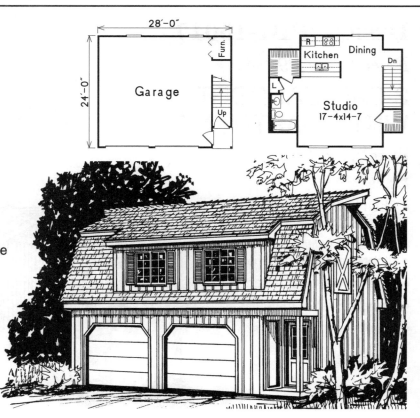

DESIGN #PB1-15027

Price Code P9

2 CAR GARAGE APARTMENT - STUDIO

- Size - 24' x 24'
- Building height - 22'-4"
- 576 square feet
- Contemporary style with private outside entrance
- Complete list of materials
- Step-by-step instructions

To order plans use the form on page 83 or call toll-free 1-800-373-2646

BUILDINGS...

53

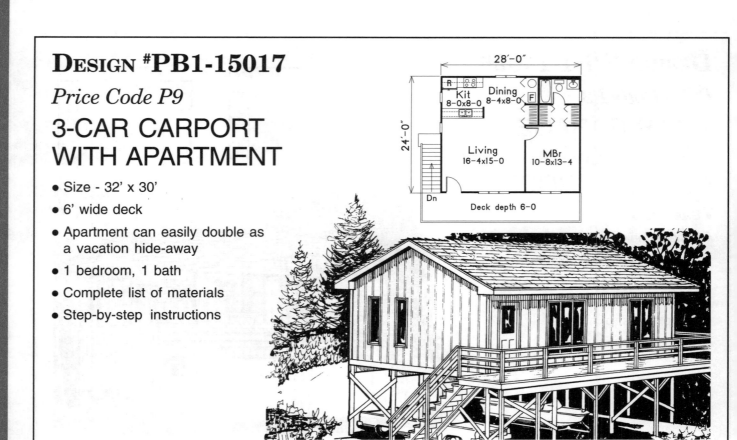

DESIGN #PB1-15017

Price Code P9

3-CAR CARPORT WITH APARTMENT

- Size - 32' x 30'
- 6' wide deck
- Apartment can easily double as a vacation hide-away
- 1 bedroom, 1 bath
- Complete list of materials
- Step-by-step instructions

28'-0"

24'-0"

R

Kit
8-0x8-0

Dining
8-4x8-0

F

Living
16-4x15-0

MBr
10-8x13-4

Dn

Deck depth 6-0

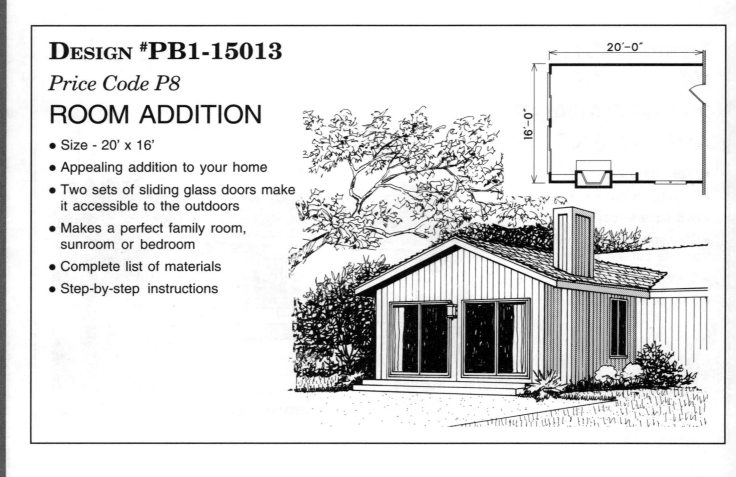

DESIGN #PB1-15013

Price Code P8

ROOM ADDITION

- Size - 20' x 16'
- Appealing addition to your home
- Two sets of sliding glass doors make it accessible to the outdoors
- Makes a perfect family room, sunroom or bedroom
- Complete list of materials
- Step-by-step instructions

20'-0"

16'-0"

To order plans use the form on page 83 or call toll-free 1-800-373-2646

BUILDINGS...

DESIGN #PB1-15005

Price Code P8

3 SEASONS ROOM

- Size - 20' x 16'-3"
- Perfect for entertaining
- Plenty of sunlight permits plants and flowers
- Complete list of materials
- Step-by-step instructions

20'-0"

16'-3"

vaulted

DESIGN #PB1-15019

Price Code P8

SUNROOM ADDITION

- Size - 16' x 16'
- Skylights brighten interior
- Sliding glass doors bring the outdoors in
- Space could also be a private home office
- Complete list of materials
- Step-by-step instructions

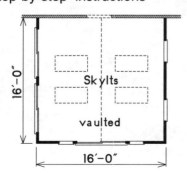

16'-0"

Skylts

vaulted

16'-0"

To order plans use the form on page 83 or call toll-free 1-800-373-2646

Design #PB1-15003

Price Code P7

SCREENED PORCH

- Size - 16'-6" x 12'-3"
- Perfect addition to any home
- Features vaulted ceiling for spaciousness
- Complete list of materials
- Step-by-step instructions

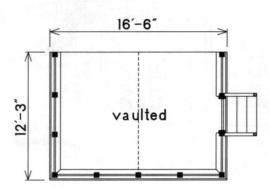

Design #PB1-15018

Price Code P7

SCREENED PORCH

- Size - 16' x 12'
- Vaulted ceiling creates spaciousness
- Inviting area for entertaining
- Adds value to your home
- Complete list of materials
- Step-by-step instructions

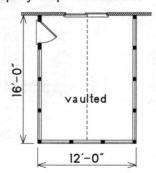

To order plans use the form on page 83 or call toll-free 1-800-373-2646

BUILDINGS...

DESIGN #PB1-15024

Price Code P7

LARGE POOLSIDE STRUCTURE

- Size - 20' x 22'
- Two dressing areas both with shower and toilet
- Covered area ideal for snack/drink bar
- Storage area accessible to outdoors for lawn and pool equipment
- Complete list of materials
- Step-by-step instructions

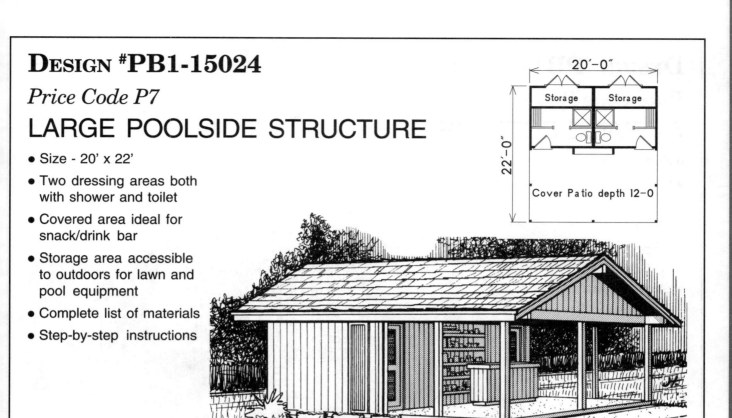

20'-0"

22'-0"

Storage   Storage

Cover Patio depth 12-0

DESIGN #PB1-15021

Price Code P7

WORKROOM WITH COVERED PORCH

- Size - 24' x 20'
- Easy access through double-door entry
- Interior enhanced by large windows
- Large enough for storage
- Complete list of materials
- Step-by-step instructions

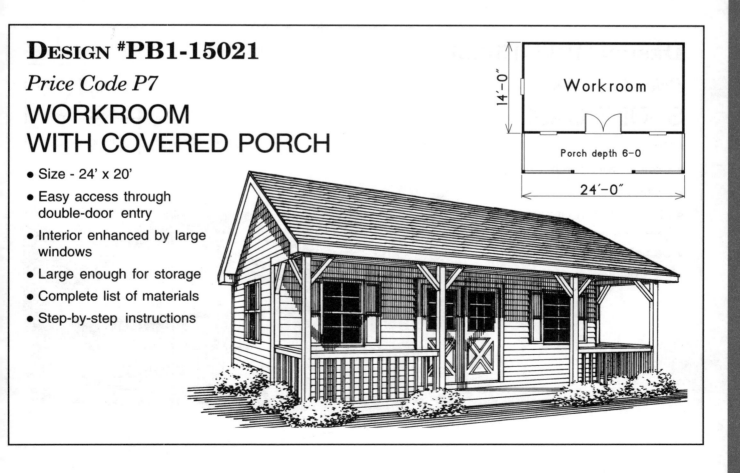

14'-0"

Workroom

Porch depth 6-0

24'-0"

BUILDINGS...

To order plans use the form on page 83 or call toll-free 1-800-373-2646

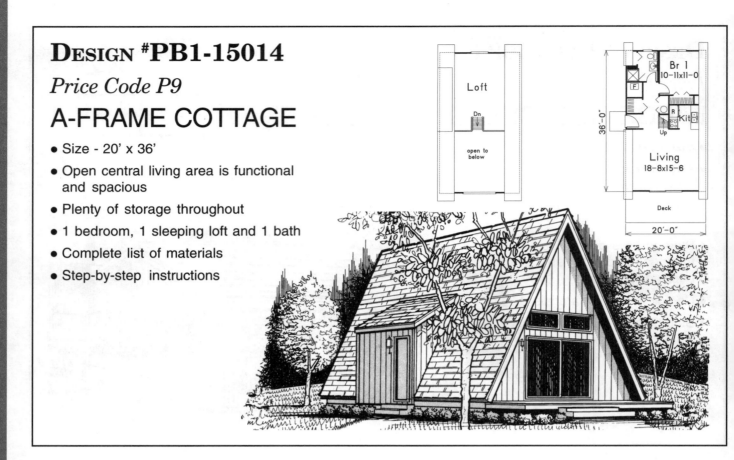

DESIGN #PB1-15014

Price Code P9

A-FRAME COTTAGE

- Size - 20' x 36'
- Open central living area is functional and spacious
- Plenty of storage throughout
- 1 bedroom, 1 sleeping loft and 1 bath
- Complete list of materials
- Step-by-step instructions

Loft

Dn

open to below

Br 1
10-11x11-0

F

R

Kit

Up

Living
18-8x15-6

36'-0"

Deck

20'-0"

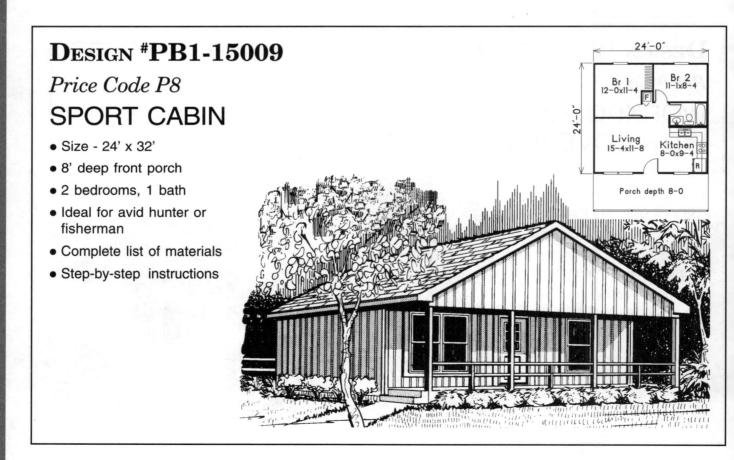

DESIGN #PB1-15009

Price Code P8

SPORT CABIN

- Size - 24' x 32'
- 8' deep front porch
- 2 bedrooms, 1 bath
- Ideal for avid hunter or fisherman
- Complete list of materials
- Step-by-step instructions

24'-0"

Br 1
12-0x11-4

Br 2
11-1x8-4

F

Living
15-4x11-8

Kitchen
8-0x9-4

R

24'-0"

Porch depth 8-0

To order plans use the form on page 83 or call toll-free 1-800-373-2646

DESIGN #PB1-15023

Price Code P8

HORSE BARN - 4 STALL

- Size - 36' x 32'
- Building height - 13'-4"
- Four box stalls with doors to covered walkway
- Includes tack room, feed storage and sliding doors
- Complete list of materials

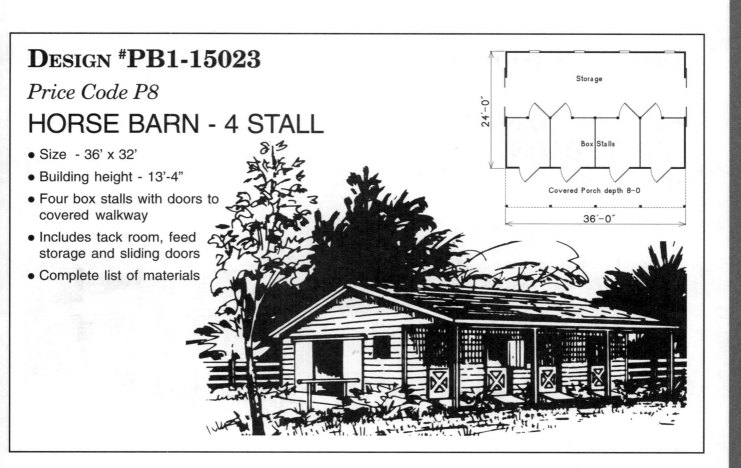

Storage

Box Stalls

Covered Porch depth 8-0

24'-0"

36'-0"

DESIGN #PB1-15022

Price Code P7

HORSE BARN - 2 STALL

- Size - 20' x 20'
- Compact, yet extra storage for feed
- Sliding side door into storage area
- Complete list of materials
- Step-by-step instructions

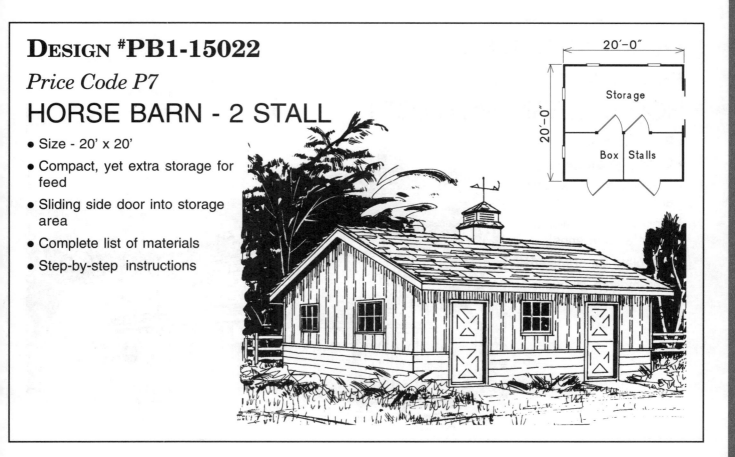

20'-0"

20'-0"

Storage

Box | Stalls

To order plans use the form on page 83 or call toll-free 1-800-373-2646

BUILDINGS...

DESIGN #PB1-15010

Price Code P8

POLE BUILDING - HORSE BARN

- Size - 36' x 32'
- Building height - 13'-9"
- Walkway connects all four stalls and leads to tack room
- Spacious feed storage area
- Sliding door on each end of structure
- Complete list of materials
- Step-by-step instructions

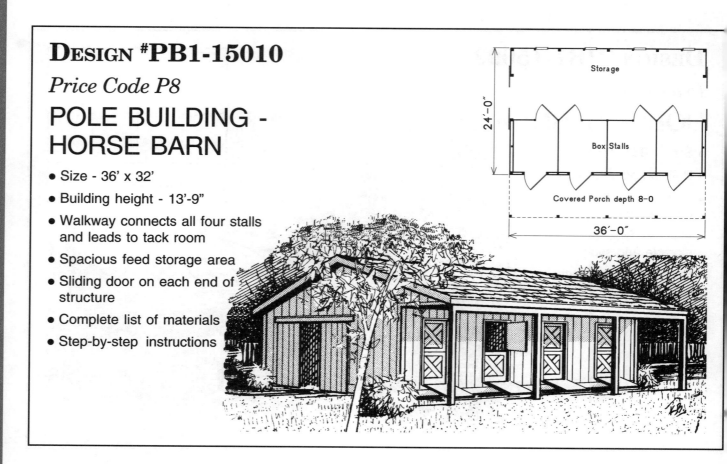

DESIGN #PB1-15012

Price Code P8

HORSE BARN WITH LOFT

- Size - 26' x 48'
- Ceiling height - 9'
- Features four box stalls and loft storage area
- Work area with sliding doors at both ends
- Complete list of materials
- Step-by-step instructions

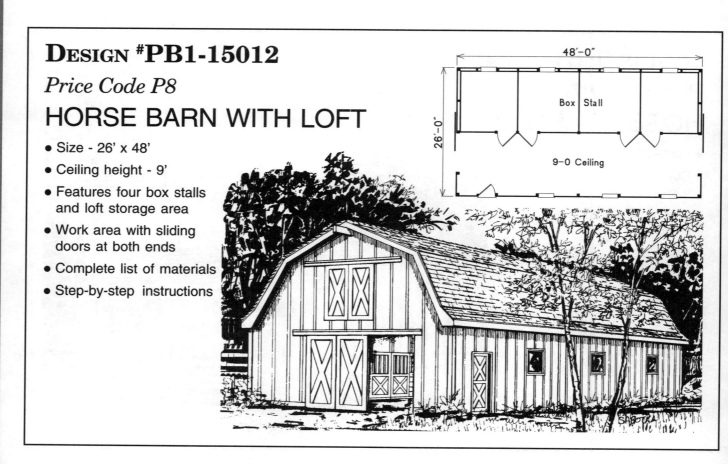

To order plans use the form on page 83 or call toll-free 1-800-373-2646

DESIGN #PB1-15002

Price Code P8

MULTI-PURPOSE BARN

- Size - 36' x 24'
- Ideal machine storage or as a three-stall horse barn
- Loft designed for 100-pound-per-square-foot live load
- Complete list of materials
- Step-by-step instructions

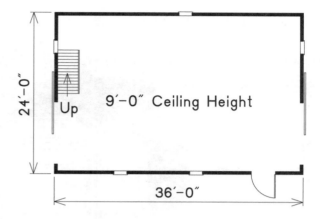

24'-0"

Up

9'-0" Ceiling Height

36'-0"

BUILDINGS...

DESIGN #PB1-15016

Price Code P8

POLE BUILDING - OPEN SHED

- Size - 36' x 13'
- Lofting storage or machinery storage
- Building can be lengthened by adding additional 12' bays
- Complete list of materials
- Step-by-step instructions

36'-0"

13'-0"

DESIGN #PB1-15006

Price Code P8

POLE BUILDING - EQUIPMENT SHED

- Size - 40' x 24'
- This design can be lengthened by adding as many 10' bays as needed
- Separated space with door perfect for workshop or storage
- Complete list of materials
- Step-by-step instructions

24'-0"

Shop Area   10'-0" Ceiling Height

40'-0"

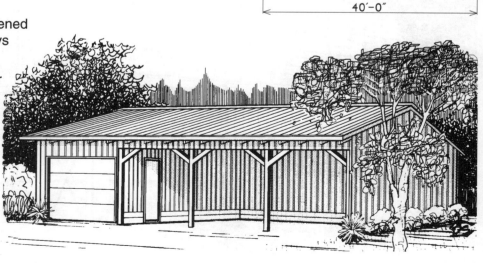

To order plans use the form on page 83 or call toll-free 1-800-373-2646

BUILDINGS...

DESIGN #PB1-15001

Price Code P8

POLE BUILDING/SHOP/GARAGE

- Size - 40' x 28'
- 14' ceiling allows room for larger equipment
- Designed for easy maintenance
- Sliding 12' x 12' side door provides easy accessibility
- Complete list of materials
- Step-by-step instructions

40'-0"

28'-0"

14'-0" Ceiling Height

DESIGN #PB1-15004

Price Code P8

POLE BUILDINGS

- 4 popular sizes

  32' x 24'

  40' x 24'

  40' x 32'

  48' x 32'

- Features either 10' or 12' ceiling heights
- Designed for easy maintenance
- Complete list of materials
- Step-by-step instructions

48'-0"

40'-0"

32'-0"

12'-0" Ceiling Height

40'-0"

32'-0"

24'-0"

10'-0" Ceiling Height

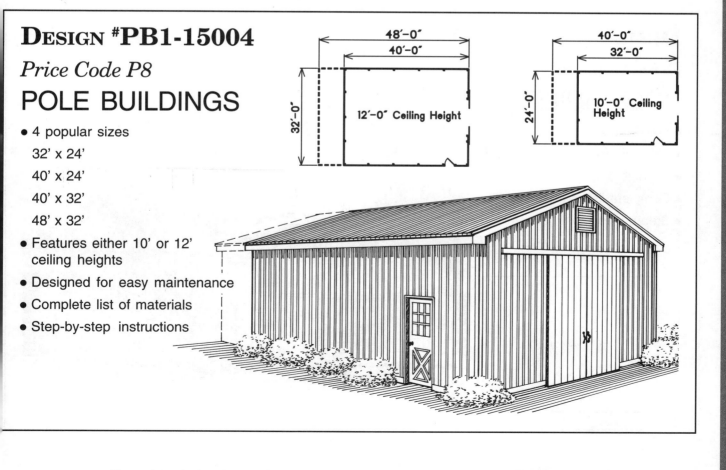

To order plans use the form on page 83 or call toll-free 1-800-373-2646

DESIGN #PB1-15008

Price Code P8

POLE BUILDING - MACHINE SHED

- Size - 40' x 64'
- Features 12' ceiling
- Sliding doors on two sides
- Complete list of materials
- Step-by-step instructions

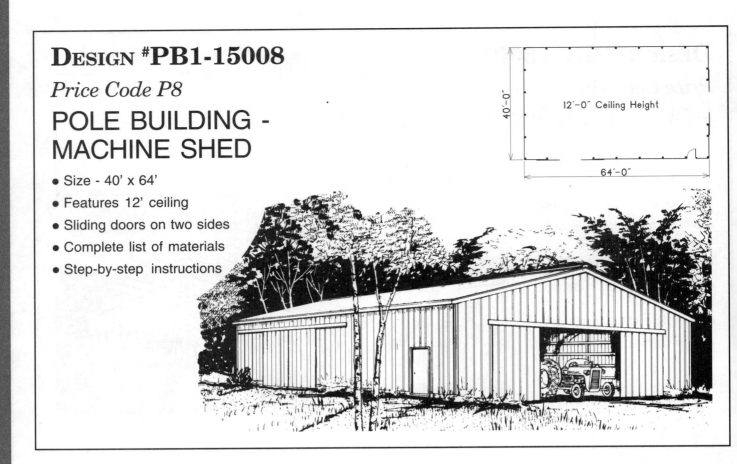

40'-0"
12'-0" Ceiling Height
64'-0"

DESIGN #PB1-15007

Price Code P8

POLE BUILDING

- Size - 32' x 40'
- 10' wide door perfect space for animal shelter
- 10' high ceiling height
- Complete list of materials
- Step-by-step instructions

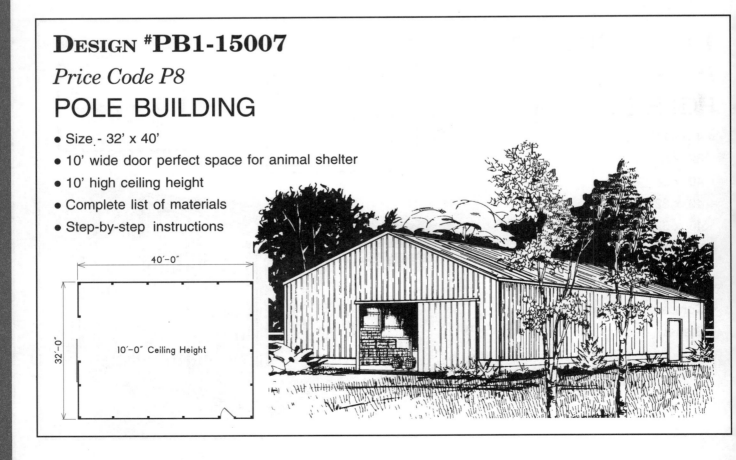

40'-0"
32'-0"
10'-0" Ceiling Height

To order plans use the form on page 83 or call toll-free 1-800-373-2646

Design #PB1-13012

Price Code P3

EASY DECKS

- 3 great sizes -
    - 8' x 12'
    - 12' x 12'
    - 16' x 12'
- Low cost construction
- Can be built with standard lumber
- Adaptable to all grades
- Complete list of materials
- Step-by-step instructions

Design #PB1-13022

Price Code P4

HEXAGON DECK

- Size - 11'-6" x 10'-0"
- Free standing design
- Atttractive deck with a choice of two railing styles
- Simple construction - easy to build
- Complete list of materials
- Step-by-step instructions

To order plans use the form on page 83 or call toll-free 1-800-373-2646

DESIGN #PB1-13006

Price Code P4

RAISED PATIO DECKS

- 2 sizes -
    - 12' x 12'
    - 16' x 12'
- Both decks can be constructed at any height
- Can be built to fit any lot situation
- Complete list of materials
- Step-by-step instructions

DESIGN #PB1-13007

Price Code P4

POOL DECK

- Size - 16' x 14'
- Can be built to fit any size pool
- Simple but sturdy design with built-in gate
- Helps cleaning and maintaining pool a breeze
- Complete list of materials
- Step-by-step instructions

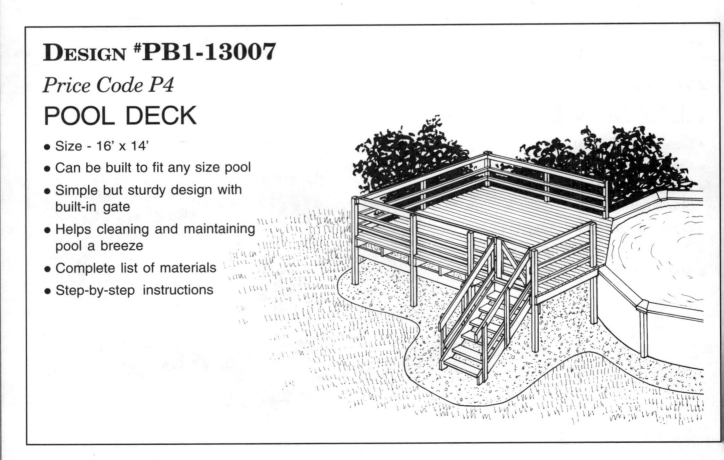

To order plans use the form on page 83 or call toll-free 1-800-373-2646

Design #PB1-13020

Price Code P5

TWO LEVEL SPA DECK

- Overall size - 20'-0" x 14'-0"
    - upper deck - 10'-9" x 11'-3"
    - lower deck - 14'-0" x 14'-0"
- Designed for self-contained portable spas
- Free standing or next to house
- Complete list of materials
- Step-by-step instructions

Design #PB1-13008

Price Code P5

TWO LEVEL DECK

- Overall size - 14' x 15'
    - lower deck - 8' x 8'
    - upper deck - 12' x 9'
- Unique, attractive deisgn features a two-level deck and bench
- Adds great value to your home
- Complete list of materials
- Step-by-step instructions

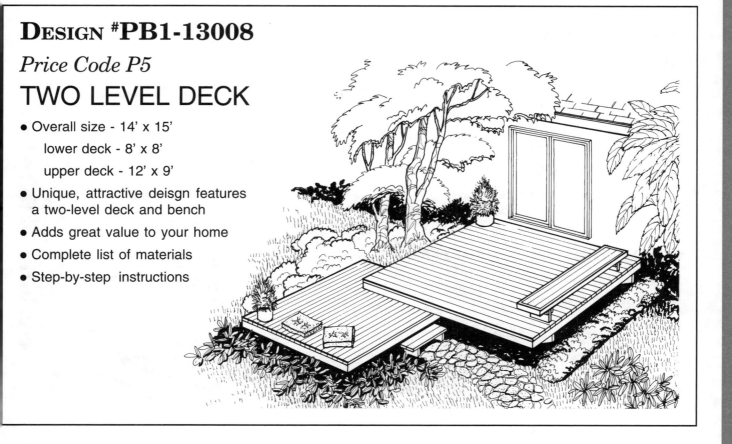

DECKS AND GAZEBOS...

To order plans use the form on page 83 or call toll-free 1-800-373-2646

DESIGN #PB1-13011

Price Code P5

HIGH-LOW DECK

- Upper deck size - 10'-0" x 8'-0"
  Lower deck size - 15'-6" x 13'-0"

- Designed as an add-on to an existing deck or as a complete unit

- Benches can be arranged as needed

- Features a unique conversation area or optional fire pit

- Complete list of materials

- Step-by-step instructions

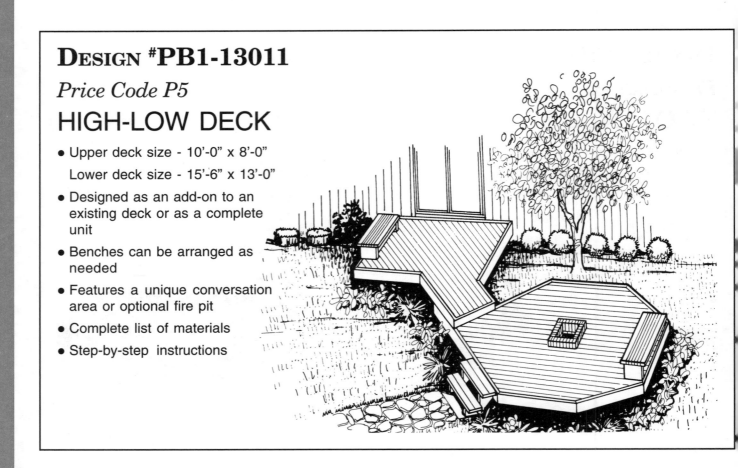

DESIGN #PB1-13010

Price Code P4

TWO LEVEL GARDEN DECK

- Overall size - 12' x 19'
  main deck - 12' x 16'
  upper deck - 8' x 8'

- Unique design features decorative plant display area or sundeck

- Built-in seating

- Can be free standing or attached

- Complete list of materials

- Step-by-step instructions

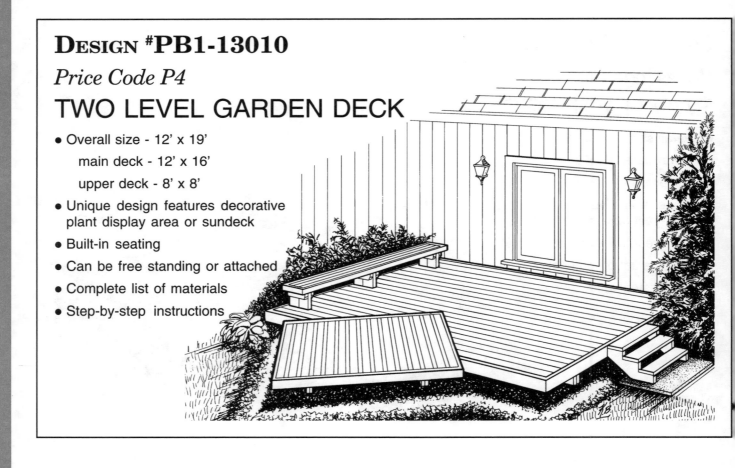

To order plans use the form on page 83 or call toll-free 1-800-373-2646

Design #PB1-13021

Price Code P5

TWO LEVEL RAISED DECK

- Overall size - 21'-0" x 24'-0"
    - upper deck - 12'-0" x 12'-9"
    - lower deck - 18'-0" x 12'-0"
- Can be built at any height
- Adaptable to any lot situation
- Complete list of materials
- Step-by-step instructions

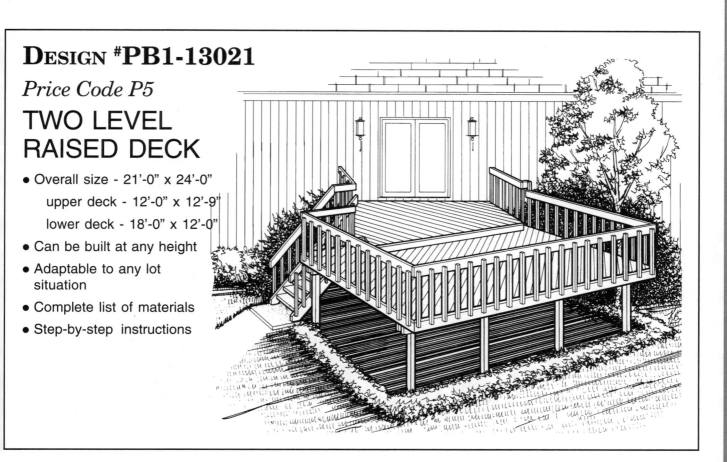

Design #PB1-13017

Price Code P5

SPLIT LEVEL DECK

- Overall size - 20' x 14'
    - upper deck - 12' x 12'
    - lower deck - 9' x 8'
- Can be built with standard lumber
- Adaptable to all grades
- Complete list of materials
- Step-by-step instructions

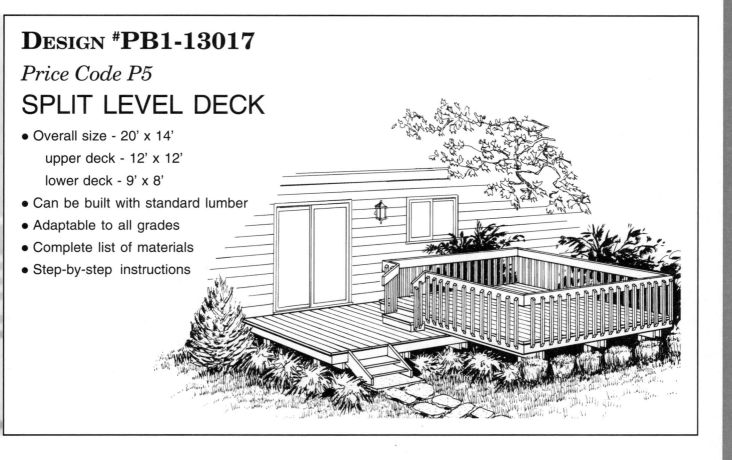

DECKS AND GAZEBOS...

To order plans use the form on page 83 or call toll-free 1-800-373-2646

DESIGN #PB1-13015

Price Code P4

PATIO COVERS

- Two sizes -
    - 12' x 12'
    - 16' x 12'
- Designed to cover an existing deck or patio or used as a pavilion
- Can be built with standard lumber
- Plan detailed with an alternate bench design
- Complete list of materials
- Step-by-step instructions

DESIGN #PB1-13009

Price Code P5

SHADED DECK

- Size - 16'-0" x 10'-0" x 9'-6" high
- Deck design has a sun screen covering
- Enhance your outdoors with this shaded deck
- Complete list of materials
- Step-by-step instructions

DESIGN #PB1-13018

Price Code P5

GARDEN ENTRYWAY

- Size - 8' x 8'
- Height peak to grade - 10'-10"
- Unique, attractive design that will complement either your garden or home
- Simple construction
- Complete list of materials
- Step-by-step instructions

DESIGN #PB1-13027

Price Code P5

FOUR-SIDED GAZEBO

- Size - 10' x 8'
- Height from top of floor to peak - 11'-0"
- Gable roof construction
- A unique and functional addition to your yard
- Adds additional shade and privacy for outside entertaining
- Complete list of materials
- Step-by-step instructions

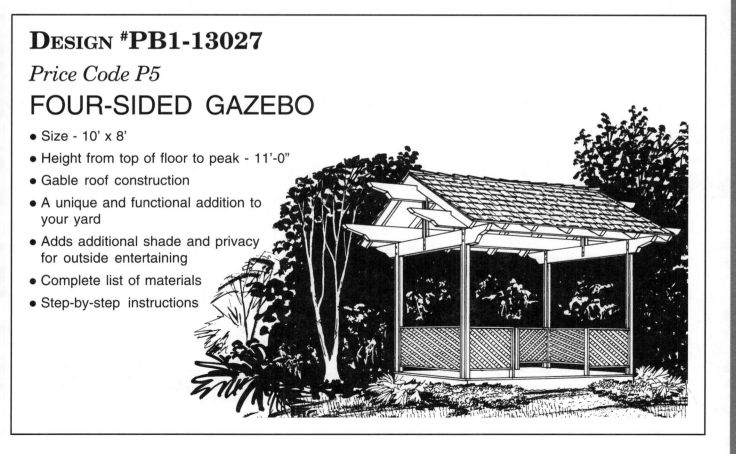

To order plans use the form on page 83 or call toll-free 1-800-373-2646

DESIGN #PB1-13019

Price Code P5

SIX-SIDED GAZEBO

- Size - 8'-3" x 9'-6"
- Height floor to peak - 12'-10"
- Complements any setting
- Cozy gazebo great for entertaining a small group
- Complete list of materials
- Step-by-step instructions

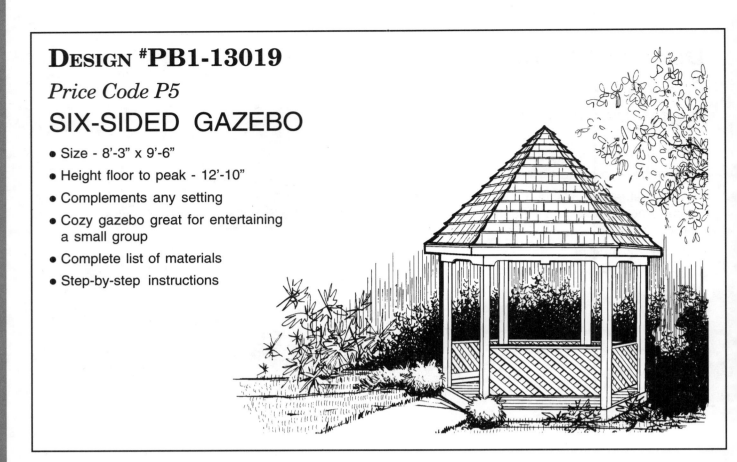

DESIGN #PB1-13031

Price Code P6

TIERED DECK WITH GAZEBO

- Sizes -

    Overall area - 28'-6" x 15'-6"
    (without gazebo)

    Deck "A" - 8'-6" x 15'-6"

    Deck "B" - 6'-0" x 8'-6"

    Deck "C" - 14'-0" x 12'-0"

    Gazebo "D" - 8'-6 sided

    Walkway "E" - 3'-0" x 7'-0"

- Gazebo offers privacy and shade
- Build complete or add on later
- Complete list of materials
- Step-by-step instructions

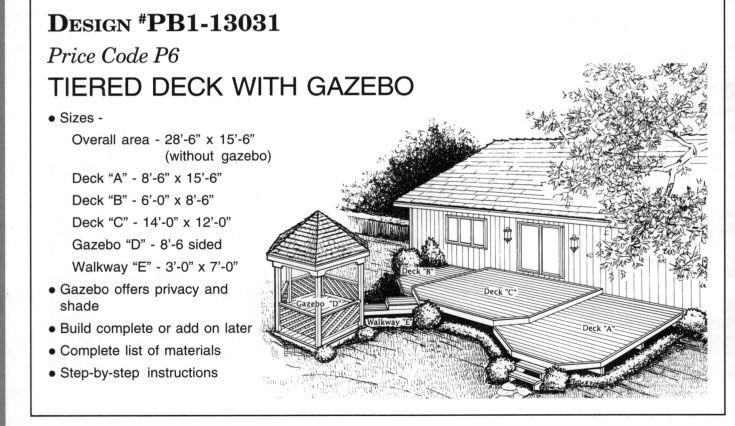

To order plans use the form on page 83 or call toll-free 1-800-373-2646

DESIGN #PB1-13026

Price Code P4

PATIO COVERS - ROOF/SUN SHADE

- Patio roof size - 16' x 9'
- Sun shade size - 20' x 10'
- A unique and functional addition to your home
- Complete list of materials
- Step-by-step instructions

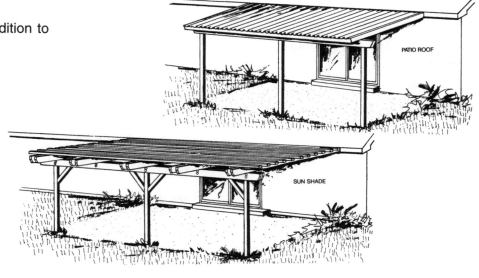

DESIGN #PB1-13013

Price Code P3

ENTRY PORCHES

- Size - 8'-0" x 5'-9"
- Two popular styles - contemporary and colonial
- Attractive designs to fit any type of home
- Can be free standing or attached
- Adaptable for trailer or home use
- Complete list of materials
- Step-by-step instructions

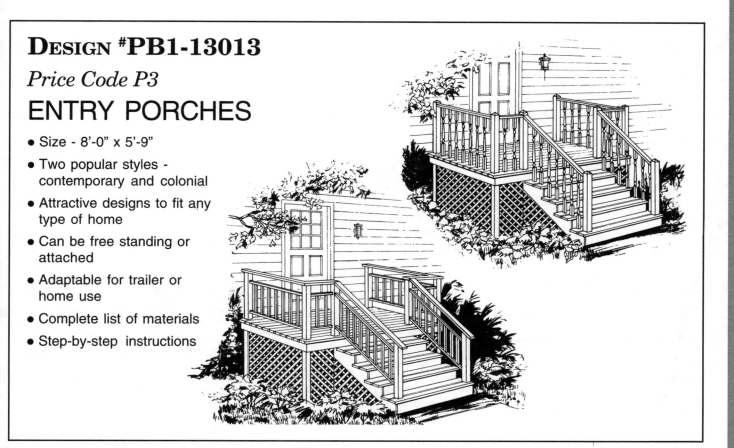

To order plans use the form on page 83 or call toll-free 1-800-373-2646

DESIGN #PB1-13023

Price Code P3

DECK RAILINGS - 5 STYLES

- Can easily be added to your existing deck
- Complement any design
- Easily adaptable to any outside structure
- Complete list of materials
- Step-by-step instructions

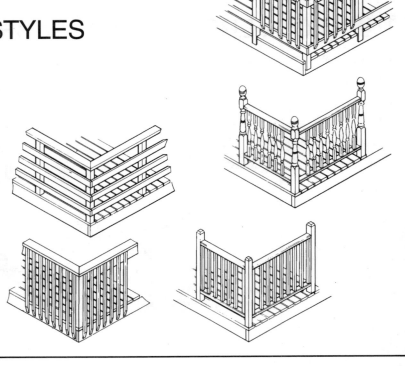

DESIGN #PB1-13014

Price Code P4

DECK ENHANCEMENTS

- Four unique designs -
  planter box size - 2'-0" x 2'-0"
  decorative screen - 7'-0" x 5'-6"
  bench - 6'-0" x 1'-8"
  end table - 2'-6" x 1'-5"
- Adds to any existing deck
- Can be free standing or attached to your deck
- Complete list of materials
- Step-by-step instructions

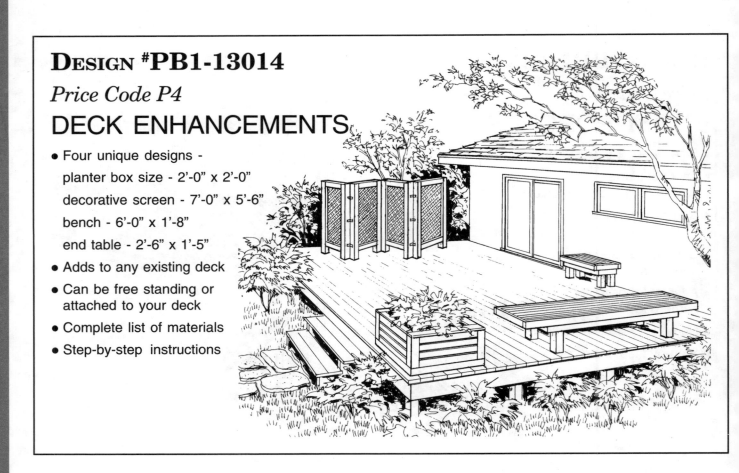

To order plans use the form on page 83 or call toll-free 1-800-373-2646

DESIGN #PB1-13001

Price Code P6

OCTAGON-GAZEBO

- Size - 11'-6" x 11'-6"
- Height floor to peak - 14'-7"
- Large gazebo has plenty of space for outside entertaining
- This attractive structure will complement any outside setting
- Complete list of materials
- Step-by-step instructions

DESIGN #PB1-13028

Price Code P5

SIX-SIDED GAZEBO

- Size - 9'-10" x 8'-8"
- Height from top of floor to peak - 10'-9"
- Ideal for small gatherings
- This traditional design will enhance any outdoor setting
- Complete list of materials
- Step-by-step instructions

To order plans use the form on page 83 or call toll-free 1-800-373-2646

DECKS AND GAZEBOS...

DESIGN #PB1-13002

Price Code P6

DECK WITH GAZEBO

- Size - 24'-0" x 15'-6"
- Height floor to peak - 12'-2"
- Perfect for outdoor entertaining
- Gazebo adds unique flair to this deck
- Complete list of materials
- Step-by-step instructions

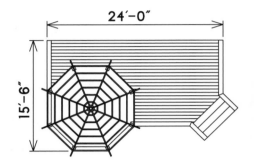

DESIGN #PB1-13029

Price Code P4

BAY DECK WITH RAILING

- Size - 20'-6" x 12'-6"
- Adds beauty and value to your home
- Unique layout with built-in bay
- Complete list of materials
- Step-by-step instructions

DECKS AND GAZEBOS...

Design #PB1-13003

Price Code P4

EXPANDABLE DECKS

- 6 popular sizes -

12' x 10'	16' x 10'	20' x 10'
12' x 12'	16' x 12'	20' x 12'

- Functional decks in a variety of sizes to fit your every need
- Complete list of materials
- Step-by-step instructions

Design #PB1-13030

Price Code P3

ENTRY PORCHES

- 2 popular styles -

    Plan 1 - 6'-5" x 5'-5"

    Plan 2 - 7'-5" x 6'-5"

- Functional porches that enhance any entrance
- Complete list of materials
- Step-by-step instructions

PLAN 1

PLAN 2

DECKS AND GAZEBOS...

To order plans use the form on page 83 or call toll-free 1-800-373-2646

DESIGN #PB1-13004

Price Code P5

3 BRIDGES

- 3 styles and sizes -
    - Plan 1 - 18'-0" x 5'-0"
    - Plan 2 - 13'-5" x 5'-0"
    - Plan 3 - 11'-0" x 5'-0"
- Enhance your outdoors
- Complete list of materials
- Step-by-step instructions

PLAN 1

PLAN 2

PLAN 3

DESIGN #PB1-13016

Price Code P4

EASY PATIO COVER

- Size - 16' x 12'
- Attractive patio cover features a sun screen covering
- Add value and beauty to your home
- Complete list of materials
- Step-by-step instructions

DESIGN #PB1-13005

Price Code P3

LOW PATIO DECKS

- 3 sizes -
    - 12' x 12'
    - 16' x 12'
    - 20' x 12'
- Built-in seating
- Perfect for entertaining
- Complete list of materials
- Step-by-step instructions

DESIGN #PB1-13024

Price Code P5

DECK WITH SUNKEN DINING AREA

- 2 sizes -
    - 18' x 18'
    - 20' x 20'
- Unique sunken area adds interest to this deck
- Perfect addition to enhance outdoor entertaining
- Complete list of materials
- Step-by-step instructions

DECKS AND GAZEBOS...

To order plans use the form on page 83 or call toll-free 1-800-373-2646

DESIGN #PB1-16003

Price Code P2

GUN/CURIO CABINET

- Size - 40" x 16" x 75" high
- Elegant and tradition in design
- Gun cabinet will hold 6 guns up to 52" long with ample storage below
- Curio cabinet will hold your cherished items on adjustable shelves and also offer plenty of storage in lower cabinet
- Complete list of materials
- Step-by-step instructions

DESIGN #PB1-16002

Price Code P2

BUNK BEDS

- Bed size - 3'-5" x 7'-1" x 5'-4" high
- Mattress size - 39" x 75"
- This bunk bed design is both versatile and durable
- Great for growing family or overnight guests
- Complete list of materials
- Step-by-step instructions

DESIGN #PB1-16005

Price Code P2

ROCKING CRADLE

- Size - 39" x 20" x 35" high
- Complements any furniture setting
- No need to custom order - 18" x 36" standard mattress will fit this attractive cradle and makes decorating easy
- Complete list of materials
- Step-by-step instructions

To order plans use the form on page 83 or call toll-free 1-800-373-2646

DESIGN #16004

Price Code P2
WORK BENCH WITH CART

- Table size - 70" x 29" x 39" high
- Cart size - 24" x 23" x 36" high
- Practical work bench has a very unique feature . . . a mobile tool caddy that rolls out from under the work bench for added workspace
- Perfect for storing tools and enough work space to handle larger projects
- Complete list of materials
- Step-by-step instructions

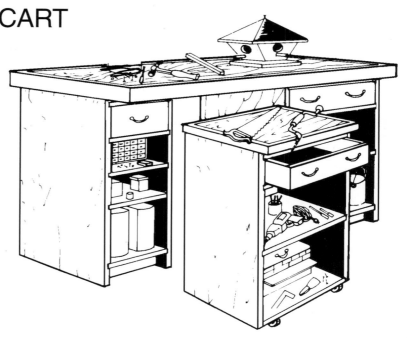

DESIGN #PB1-16001

Price Code P2
WET BAR

- Size - 8' x 6' x 3'-6" high
- Attractive serving bar perfect for any room
- Plenty of storage area
- Ideal design for entertaining
- Complete list of materials
- Step-by-step instructions

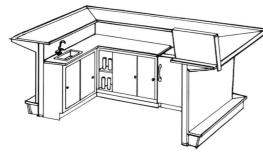

INDOOR...

To order plans use the form on page 83 or call toll-free 1-800-373-2646

PROJECT PLANS INDEX